BASIC ALGEBRA AND GEOMETRY MADE A BIT EASIER LESSON PLANS

Also by Larry Zafran

WEIGHT LOSS MADE A BIT EASIER:
Realistic and Practical Advice for Healthy Eating and Exercise

A REALISTIC EATING AND EXERCISE RECORD BOOK:
A Six-Month Weight Loss Log and Journal
for Dedicated Individuals

THE COMPLETE MUSIC PRACTICE RECORD BOOK:
A Six-Month Log and Journal for Dedicated Students

AMERICA'S (MATH) EDUCATION CRISIS:
Why We Have It and Why We Can('t) Fix It

THE REGIFTABLE GIFT BOOK:
The Gift That Keeps On Regiving

MATH MADE A BIT EASIER:
Basic Math Explained in Plain English

MATH MADE A BIT EASIER WORKBOOK:
Practice Exercises, Self-Tests, and Review

MATH MADE A BIT EASIER LESSON PLANS:
A Guide for Tutors, Parents, and Homeschoolers

BASIC ALGEBRA AND GEOMETRY MADE A BIT EASIER:
Concepts Explained in Plain English, Practice Exercises, Self-Tests

www.LarryZafran.com

BASIC ALGEBRA AND GEOMETRY MADE A BIT EASIER LESSON PLANS

A Guide for Tutors, Parents, and Homeschoolers

LARRY ZAFRAN

Self-published by Author

BASIC ALGEBRA AND GEOMETRY MADE A BIT EASIER
LESSON PLANS: A Guide for Tutors, Parents, and Homeschoolers

Self published by Author

Book design by Larry Zafran

Printed in the United States of America
First Edition printing May 2010
First Edition revised May 2011

ISBN-10: 1-4515-5820-1
ISBN-13: 978-1-45-155820-3

Please visit the companion website below for additional information, to ask questions about the material, to provide feedback, or to contact the author for any purpose.

www.MathWithLarry.com

CONTENTS

CHAPTER ZERO

INTRODUCTION

ABOUT THE *MATH MADE A BIT EASIER* SERIES

This is the fifth book in the self-published *Math Made a Bit Easier* series which will be comprised of at least eight books. The goal of the series is to explain math "in plain English" as noted in the subtitle of the first book.

The first book in the series covers basic math which is the foundation of this book and all later math. It is essential that the first book not be skipped, and that it is thoroughly mastered before working with this book. Like all the books in the series, it is available for free reading in its entirety on Google Books for those who cannot or choose not to purchase it.

The second book in the series is a companion workbook of review, practice exercises, and self-tests which corresponds directly to the first book. The third book is a companion lesson plan book for the first book. The fourth book (which is the companion to this book) covers basic algebra and geometry. With that material mastered, the student should be well-prepared for standardized exams such as the SAT or GED, with the exception of a few more advanced topics which will be covered in the next set of books.

A later book in the series will serve as "crash course" of material, intended only for students who, through poor planning, have left with themselves with no better option. An auxiliary book will discuss America's (math) education crisis.

In addition to providing math instruction, the series also attempts to explain the truth about why students struggle with math, and what can be done to remedy the situation. As a totally independent and self-published author, I am able to write with the necessary candidness.

Unlike many commercial math books, the series does not imply that learning math is fast, fun, or easy. It requires time and effort on the part of the student. It also requires that the student be able to remain humble as s/he uncovers and fills in all of his/her math gaps, and that s/he makes time to do so.

THE PURPOSE AND TARGET AUDIENCE OF THIS BOOK

This book is intended for tutors and/or parents who provide math instruction for public, private, or homeschooled students. It contains 50 lesson plans which correspond directly to the fourth book of the series. As such, the reader should refer closely to the fourth book while working with this one.

This book assumes that you as the instructor have a strong familiarity and comfort level with the material, and that you have basic skills in presentation. You don't necessarily have to be a professional teacher with years of experience, but keep in mind that your students will pick up on any doubt or insecurity on your part. It is also important that you have a good understanding of how all of the topics relate to one another.

It is important to understand that for the sake of conciseness, these lesson plans are highly abridged. They are intended as an overall framework and outline which the instructor can follow, or as a source of ideas. The lessons include advice on which concepts students tend to find easy or difficult, along with suggestions on which ones require reemphasis.

Like the rest of the series, this book is not intended to be used with students who are pursuing math-related degrees, or who want to explore the richness of the subject. The lessons were designed from the realistic standpoint that most students are just trying desperately to fulfill their math goals and requirements, and have very little time in which to do so.

HOW TO USE THIS BOOK

This book directly corresponds to the chapters of the fourth book. Refer to that book in tandem with this one, using either a purchased hardcopy or the free Google Books version.

Assuming that the student has mastered all of the prerequisite material from the basic math books of the series, start this book at Lesson 1 regardless of what you believe the student's level is, or what level s/he is according to government tests, or what level s/he (or the parents) says or thinks s/he is.

It is important to spend the appropriate amount of time on each lesson—no more and no less. This amount of time will vary for each lesson, which means that you as the instructor will have to be flexible and adaptive. If a student has truly internalized the material in a lesson, it may only be necessary to spend a few minutes on it, just for the sake of assessment and review. If a lesson involves totally new material, or material that the student finds confusing, it may require one or more hours, even in the context of private tutoring. Review previous lessons as needed, especially ones which contain prerequisite knowledge for the current lesson.

THE FORMAT AND METHODOLOGY OF THE LESSONS

This book takes a standard, traditional approach to teaching math. Each lesson includes an aim or topic list, connections to earlier and/or later material, questions or problems to serve as a warm-up and motivation, demonstrative examples, points to elicit, and practice exercises which in most cases reference the fourth book of the series. Some of the lessons also include ideas for embellishment and/or points to reemphasize.

The warm-ups in this book are based on the philosophy that there is no harm in presenting the student with a problem which s/he may have trouble solving. In so doing, the student is forced to think about the problem and apply earlier concepts in an effort to solve it. This technique also provides the instructor with guidance as to how best to begin the lesson. Never allow a student to give up prematurely. The warm-ups are not tests which will be graded. They should be presented simply as problems for brainstorming and discussion.

These lessons may be "dry" in compared to curricula that utilize non-traditional methods for teaching math. As I explain in the first book, most educators are very opposed to such special methods. It is important to understand that when a student takes a standardized exam, s/he is expected to solve problems quickly, and on his/her own. While this book does make reference to diagrams and other techniques of instructional merit, the student must not be dependent upon them.

THE BOOK'S POSITION ON CALCULATOR USE

As described in the first book, this book takes a modern and practical view on calculator use. All but the very youngest students are typically permitted to use a calculator in class, for homework, and on exams. The trend in math education is for students to learn and apply concepts, and creatively make connections between them—it is not to see if students are able to do multi-step arithmetic by hand.

While working through the lessons, it is suggested that you adopt this practical view on calculator use. It is the concepts and procedures which are important—not the actual "number crunching." With that said, it is truly not practical for the student to have to reach for a calculator for every single minor calculation. Be especially careful with signed number arithmetic. Many students do not fully understand how to correctly input signed numbers into their calculator, and most basic calculators do not process signed numbers in the way in which the student expects.

WHY ARE ALGEBRA AND GEOMETRY HARD?

The basic algebra and geometry topics presented in this book are truly not "hard." It may be fair to describe some of the concepts as abstract, but they are not at all the "big deal" which most students make them out to be. In almost all cases, the reason why students proclaim that basic algebra and geometry are hard is because they never fully learned the basic math concepts upon which those topics depend.

Most of the lessons in this book outline the prerequisite basic math skills required for the lesson. If the student does not have those prerequisite skills, you as the instructor have a choice. You must choose between muddling through the lesson, leaving the student (and his/her parents) with the impression that s/he understands it, or, you can explain the situation to the student (and his/her parents), and teach the required prerequisite topics before teaching the algebra and geometry topics which build upon them. The latter option leads to success. The former leads to continued failure and frustration.

LOGISTICAL ASPECTS OF TUTORING

To be an effective instructor, you alone must control the flow of instruction, with the extremely rare exception in the case of a highly mature student who is only very slightly behind in math. This is counterintuitive because good customer service dictates that you let the student (or his/her parents) tell you what topics to cover, and in what order, and with what time constraints. While adopting such a policy will likely result in the student (and his/her parents) considering you to be an outstanding and attentive instructor, in most cases it is pure fraud which is being perpetuated.

Your methodology should be the same for every student, although it will be implemented differently for each. It is as simple as identifying the student's difficulties, tracing them back to their source, remedying the situation starting from the source, and moving forward only when the student is ready.

Admittedly this is very difficult if the student says, "I need help with tonight's homework," or "I have to pass an important test tomorrow," or similar. It is also difficult if the student derails the entire session by pointing to a very challenging problem in the book (perhaps one that is labeled as such, and was not even assigned), but insists that you "show" him/her how to do it, since s/he doesn't "get it."

It is important to understand that most students will do anything in their power to avoid receiving remedial instruction. In some cases this is due to embarrassment, and in some it is because the student legitimately believes that s/he knows the material, and even has high test scores and a diploma from his/her primary and middle schools to "prove" it.

ELICITING VERSUS LECTURING

Each lesson in this book includes a "Points to Elicit" section. To be an effective instructor, try to conduct as much of your lesson as possible in the form of questions, as opposed to a lecture. You must constantly force your students to think, and to make connections between topics. Avoid stating anything yourself which the student could have stated or conjectured him/herself. Most students hate this method, but that is how learning takes place. More importantly, that is how retention takes place. Do not rob the student of the chance to think for him/herself since s/he will be on his/her own during exams.

HOW TO GET HELP WITH ANY OF THE LESSONS

I provide full support for this and the other books in the series via my comprehensive website at **www.MathWithLarry.com**. Feel free to e-mail me with any questions or comments that you have. I am happy to provide guidance on how to work with a student who is struggling with a particular topic. I can also provide additional practice exercises or demonstrative examples for any topic upon request, and of course I can answer any questions that you have about the material.

CHAPTER ONE

Lesson Plans for "Working with Algebraic Expressions"

Lessons 1 to 3

Topics Covered in This Chapter:

What is algebra?; Four ways to represent multiplication; Evaluating expressions; Algebra definitions; Combining like terms; Degree of a term; Implied coefficient of 1; Subtraction sign vs. negative sign

LESSON 1

AIM/TOPIC(S): What is algebra?; Evaluating simple algebraic expressions; Four ways to represent multiplication

CONNECTIONS TO EARLIER/LATER MATERIAL:
It is absolutely essential for the student to have fully mastered all of the material from the first book in this series (Basic Math). This lesson begins the student's study of algebra.

PREREQUISITE KNOWLEDGE: The second and fourth book in this series contain self-tests of prerequisite material which should be administered to the student. Do not proceed with this book unless the student can handle the tests with confidence, since doing so will result in the student being "lost."

WARM-UP: Present the problem ? – 7 = 2 which the student should find easy. Ask him/her how s/he got the answer. Ask when s/he worked with problems like that, to which s/he will likely say one of the lower grades. Explain that s/he was really working with basic algebra, which at a basic level is the study of how to deduce and work with unknown values. Use this to lead into the lesson.

MOTIVATION: Algebra is a vast field of study in math. If the student asks, "Why do we have to study it?," explain that a large variety of problems can be solved using algebra. Algebra forms the basis of later math which the student will be tested on, and which s/he may choose to pursue further. Most students who are comfortable with basic math tend to find algebra fun and interesting. It is the students who never mastered the prerequisite material who struggle.

POINTS TO ELICIT: Explore the ways in which the warm-up exercise can be solved, and the fact that there was only one correct answer. There was nothing special about using a box to represent the unknown value. We could have used a question mark or other symbol, or even a letter of the alphabet.

In algebra the warm-up exercise would most likely be written as $x - 7 = 2$, but any letter can be used. This exercise involves what is called an equation because of the equals sign. We must find the value of x such that the left side equals the right.

DEMONSTRATIVE EXAMPLES: Present the problem: "Evaluate $x + 3$ when x is 5." We could have used ";" in place of "when," and "=" in place of "is." Elicit the difference between this problem and the previous. This problem involves an expression and not an equation. Explain how the question-maker was free to choose any value of x that s/he wanted. Our task was just to substitute ("plug in") the given value into the expression, and see what it works out to be.

Demonstrate the three new ways of representing multiplication which are needed to avoid confusion since we commonly use x in algebra. Examples are 3x, 3·x, and (3)(x). Be certain to stress that the middle dot is not a decimal point, and that 3x means multiplication because of the absence of an operation symbol—not because the x looks like a multiplication sign.

PRACTICE EXERCISES: Since equations are covered in upcoming lessons, spend time evaluating a variety of simple expressions using a variety of values. Be sure to include values which are negative, positive, and 0.

INSTRUCTOR'S NOTES:

LESSON 2

AIM/TOPIC(S): Algebra definitions; Combining like terms; The degree of a term; Implied coefficients of 1; Evaluating more complicated algebraic expressions

WARM-UP: Elicit the student's thoughts on simplifying $3x + 2x + 5x$. If the student is hesitant, repeat the problem, but substituting "apples" in place of each x. Now elicit the student's thoughts on simplifying $2x + x + 4x$, and again, if necessary, substitute "apples" in place of each x, with the middle term becoming "an apple."

DEFINITIONS: In algebra, an isolated numeric value is called a constant. Elicit some examples including positives, negatives, fractions, etc. A letter representing an unknown value is called a variable. The concept is elaborated on in a later lesson.

When a numeric value is multiplying a variable, the numeric value is referred to as the coefficient of the variable. Elicit that we handled the warm-up problems by adding the coefficients. We call that procedure "combining like terms."

We define a term as a constant, a variable, or the product and quotient of constants and variables. Multi-term expressions are formed by combining different terms using addition and subtraction. Elicit examples of terms and expressions.

MOTIVATION: Revisit the first warm up question. Ask the student to evaluate it when x is 7. Elicit that it would be much easier if we had first combined like terms, although we get the same answer either way. Revisit the second warm up equation. Ask the student to evaluate when x is 8. Elicit how much easier our task would be if we first combined like terms.

POINTS TO ELICIT: The isolated x in the second warm up question was really 1x. It has an implied coefficient of 1. Some students like to call this an "invisible 1." Discuss why the implied coefficient is 1 and not 0.

DEMONSTRATIVE EXAMPLE: Present the expression $7x^2 + 4x + 3 + 8 + 9x^2 + 6x$. Elicit that if we were asked to evaluate this expression for a given value of x, it would be tedious. Instead, we can first simplify the expression by combining like terms. Most students feel comfortable with the idea that x^2 terms match x^2 squared terms, x terms match x terms, and constants match constants.

Ensure the student understands that x^2 and x terms can never be combined no matter how tempting it may look. An example would be the expression $9x^2 + 9x + 9$ which cannot be combined because it contains three unlike terms. The fact that they all involve 9 does not make the terms like.

Define the degree of a term as the exponent involved. (In later math, that definition is somewhat expanded). The degree of $7x^3$ is 3. Elicit that the degree of 8x is 1 since x^1 and x are equal. Elicit that the degree of a constant such as 4 is actually 0 because 4 could be rewritten as $4x^0$. For the purposes of combining, terms are considered like if have the same degree.

PRACTICE EXERCISES: Have the student practice combining like terms in various expressions. For now, just use expressions involving addition of positive terms.

INSTRUCTOR'S NOTES:

LESSON 3

AIM/TOPIC(S): Subtraction vs. negative signs

PREREQUISITE KNOWLEDGE: Before starting this lesson, it is absolutely essential that the student have no difficulty at all with signed number arithmetic. Ask him/her to compute 4 – 7. If s/he answers incorrectly, or displays even the tiniest bit of hesitation or doubt, there is truly no point in starting this lesson. Review the first book if necessary.

WARM-UP: Ask the student why 4 – 7 and 4 + (-7) are equivalent computations, and if the dash between the 4 and the 7 is a negative or a subtraction sign. Have the student try combining like terms in the expression $2x^2 + 6x - 3 - 9x^2 - 10x + 5$. Use any answers to lead into the lesson.

MOTIVATION: Signed number arithmetic is used constantly in algebra. Explain to the student (and his/her parents if applicable) that this lesson will not end until all of its concepts are fully mastered. It will take as little or as much time as needed.

POINTS TO ELICIT: Revisit the matter of whether the dash between the 4 and the 7 in (4 – 7) is a subtraction sign or a negative sign. Most students have tremendous difficulty with this concept. Elicit that if we want to treat the dash as negative sign for the 7 (which is fine), we will be left with two isolated values sitting side by side, namely 4 and -7. This is typically not of concern to students who will simply ask, "So what?"

Unless we are dealing with a list or a set of numbers, two values must always be connected via some operation. At this point, elicit that if we treat the dash a negative sign for the 7, we must then insert a plus sign (an implied addition operation) between the two values. This takes us back to the concept that 4 – 7 and 4 + (-7) are the same computation, which the student should have mastered in earlier work. Elicit that we could also have chosen to treat the dash between the 4 and the 7 as a subtraction operation, in which case we are dealing with the equivalent "4 minus 7."

DEMONSTRATIVE EXAMPLE: Revisit the warm-up task of combining like terms. The previous lesson covered the general concept, so the task now is to properly handle the terms to be subtracted (or treated as negatives). For the squared terms we have 2 minus 9 (or 2 + -9), giving us $-7x^2$. For the x terms we similarly get -4x, and for the constants we get positive 2. Again, this task is not possible if the student is unable to do signed number arithmetic in a non-algebra context.

For our final answer we must represent the like terms we ended up with, $-7x^2$, -4x, and 2, but of course we cannot just present those values in list format. Elicit that we can insert implied plus signs between each term as described earlier. That would give us $-7x^2$ + -4x +2. While that is not wrong, it would be customary to remove the addition sign to the left of the -4x, and treat the dash as a subtraction operation giving us a final version of -7$x^2 - 4x + 2$. Notice how the negative 7 is represented as such since it is the first term of the expression.

PRACTICE EXERCISES: Make up similar practice exercises as needed. Ensure that the student is not just doing endless rote computations, but that s/he truly understands the concept.

INSTRUCTOR'S NOTES:

CHAPTER TWO

Lesson Plans for "Solving Basic Algebraic Equations"

Lessons 4 to 6

Topics Covered in This Chapter:

Introduction to algebraic equations; General procedure and golden rule for solving algebraic equations; Equations involving the four basic operations; Equations involving decimals and fractions

LESSON 4

AIM/TOPIC(S): Introduction to algebraic equations; General procedure and golden rule for solving algebraic equations

PREREQUISITE KNOWLEDGE: Review the three previous lessons. Don't start this lesson until they are fully mastered.

WARM-UP: Present the equation $x + 5 = 8$. Ask the student to "solve for x." S/he will likely answer 3. Then present the equation $y - 34 = 97$, and ask the student to "solve for y," or at least comment on the problem. Repeat the above for the equation $z - 9 = -7$. Use any responses to lead into the lesson.

MOTIVATION: Previously we worked with algebraic expressions. There was nothing to "solve for." If we were told what value to use for a variable, we could substitute it and evaluate.

In the first warm-up problem we worked with an equation which in this case was an expression set equal to a value. We (or the question-maker) cannot just pick an arbitrary value for the variable. Our task is to determine what value of the variable will make the equation hold true. In this case the value was 3, because when x is 3, both the left side and the right side are equal to each other (both evaluate to 8).

POINTS TO ELICIT: Revisit the second problem. It is hard to solve in one's head. In the third problem, even though we are dealing with smaller numbers, the presence of a negative can lead to confusion. In order to solve algebraic equations like these (and more complicated ones later), we must have a specific procedure to follow. Algebra offers such procedures which will always give us the correct answer when followed.

DEMONSTRATIVE EXAMPLES: Revisit the first warm-up problem. Acknowledge that it was solvable mentally, but many problems are not. Elicit that we could use what we know about "fact families" to get our answer by computing 8 – 5.

Present the "golden rule" of algebra which is that we can solve for variables by using inverse operations to "undo" what we see being done to them. In this case we can see that 5 is being added to x. We know subtraction is the inverse of addition, so we can subtract 5 from the left side in order to "cancel" or "get rid of" the +5. Elicit that what is really happening is that we have -5 + 5 which is 0 (the identity element in addition).

Stress that we can't just stop there. Many students have a hard time understanding why. If we subtract 5 from only the left side, we will be left with x = 8. Have the student check the answer by substituting it in the original equation where it obviously won't work. Don't allow the student to maintain a "so what" attitude. Do not move on until s/he is convinced that 8 cannot possibly be the correct value of x in this equation.

Present the concept of an old-fashioned balance scale. To maintain the balance, we can do whatever we want to one side, as long as we do the same to the other. In this case, since we subtracted 5 from the left, we must do the same on the right. That gives us x = 3, the correct answer.

POINTS TO REEMPHASIZE: For now, just review the concepts presented. The next several lessons instruct how to apply them in various equations.

INSTRUCTOR'S NOTES:

LESSON 5

AIM/TOPIC(S): Solving one-step algebraic equations involving the four basic operations; Checking answers

WARM-UP: Present the equation $x + 7 = 2$. Have the student solve it using the procedure and "golden rule" from the last lesson. Do not allow him/her to take repeated guesses. Then present $x - 5 = 6$, and see if s/he can correctly apply what was learned. Then present $3x = -12$, and $\frac{x}{4} = 8$, and see how the student does. Use any responses to lead into the lesson.

MOTIVATION: The student will likely be pleased to know that at a basic level, this lesson is all algebra is. We follow specific steps, and we end up with the answer. We also have a means of checking our answers to ensure that they are correct.

POINTS TO ELICIT: Even though we're using x in each warm-up problem, the x will likely always take on a different value. We could have used any letter at all, and it is just a coincidence if x works out to be the same value in more than one problem.

Revisit the first warm up problem. Most students will blurt out, "5, -5, 9, -9," realizing that the answer is likely one of those possibilities. Do not permit this. Elicit that we should use the procedure from the previous lesson to get our answer. Since the expression on the left adds 7 to x, we must "undo" that using the inverse operation of subtraction. Review that whatever we do to one side, we must do to the other.

Demonstrate the traditional way of solving by writing -7 underneath the 7 and the 2. Demonstrate that while we typically draw a line through the sevens and say they "cancel," we're really computing 7 – 7 which is 0, and need not be written. On the right we have 2 – 7 which is -5. x = -5.

Explain that sometimes the variable will be a negative integer, or a fraction, or a decimal. If the student has difficulties with the signed number arithmetic, that material must be mastered before proceeding. Elicit that we can check our answer by substituting it back into the original equation. (-5) + 7 = 2. Both sides evaluate to 2, so our answer was correct.

The next three warm-up problems are solved similarly. For the second problem, since 5 is being subtracted from x, we must add 5 to both sides. We get x = 11. Check as described. For the third problem we must remember that division is the inverse operation of multiplication. Since x is being multiplied by 3, we must divide by 3 to "get rid of" the x. Elicit that the x's don't disappear. We are dealing with 3/3 which is 1, the identity element in multiplication which need not be written. After dividing both sides by 3 we get x = -4. Again, ensure that the signed number arithmetic is not a point of concern.

For the fourth problem, since x is being divided by 4, we must do the inverse operation which is to multiply both sides by 4. Demonstrate that the 4 is usually written as a large number to the left of the fraction, but it is really 4/1 which allows us to "cancel" it with the 4 in the denominator. We get x = 32. Be certain to have the student check all answers as described.

PRACTICE EXERCISES: The main book has some practice exercises which you can supplement as needed. Again, always favor comprehension of the concepts instead of rote drilling.

INSTRUCTOR'S NOTES:

LESSON 6

AIM/TOPIC(S): Solving one-step algebraic equations involving fractions and decimals

CONNECTIONS TO EARLIER/LATER MATERIAL:
This short lesson reviews the previous lessons in the book, as well as basic arithmetic involving decimals and fractions.

PREREQUISITE KNOWLEDGE: It is essential that the student have mastered the concepts from the previous lesson, and have no difficulty performing basic operations with decimals and fractions. Do not attempt this lesson if that is not the case.

CALCULATOR USE: This book acknowledges that many students are permitted to use calculators for their coursework and exams. It may be difficult to convince the student to handle algebraic computations such as the ones in this lesson by hand. Try to convince him/her that if s/he uses a calculator, it is even more important to check all of his/her answers to ensure that an error in calculator input has not been made.

WARM-UP: Present the equation $x + 1.5 = 7.24$, and ask the student to solve for x. Have him/her do the same for the equation $x - \frac{1}{6} = \frac{2}{3}$. Use any responses to lead into the lesson.

MOTIVATION: Again, this lesson serves as a good review of the algebra learned thus far. All it does it take those concepts and apply them to equations involving decimals and fractions. Many students have a hard time accepting that the algebraic procedures that they learned also work in equations which involve non-integer values. Convince the student that there is no reason to panic which is a common response.

POINTS TO ELICIT: Revisit the first warm-up problem. We learned that the procedure would be to subtract 1.5 from each side, giving us x = 5.74. We can verify the answer by following the checking procedure we learned. If the student has difficulty with this task, it is essential to determine if it is related to the algebra, or the decimal arithmetic. Obviously review the related material accordingly.

Revisit the second warm-up problem. We learned that the procedure would be to add $\frac{1}{6}$ to each side, giving us $x = \frac{5}{6}$. We can verify the answer by following the checking procedure we learned. Again, if the student struggles, determine if the problem is related to the algebra or the fraction arithmetic.

DEMONSTRATIVE EXAMPLES: Most algebraic equations involving decimals and fractions require addition or subtraction to solve. The point of the unit is to have the student not feel "locked into" integers. Depending on the student, consider supplementing the lesson with fraction and decimal one-step equations which require multiplication or division to solve.

PRACTICE EXERCISES: The main book has some practice exercises which you can supplement as needed. Again, always favor comprehension of the concepts instead of rote drilling.

POINTS TO REEMPHASIZE: Since this lesson is short, and upcoming lessons introduce more advanced algebraic procedures, consider using any extra time to thoroughly review all of the material presented thus far. Ensure the student understands that the rest of this book will build upon the algebraic techniques presented in this unit.

INSTRUCTOR'S NOTES:

CHAPTER THREE

Lesson Plans for "More Complicated Algebraic Equations"

Lessons 7 to 14

Topics Covered in This Chapter:

Introduction to variables; Simplifying algebraic equations by combining like terms; Two-step equations; Equations with variables and constants on both sides; Finding an unknown value which results in a given mean; Equations involving a variable times a fraction; Simple interest formula; Proportion problems; Distributing a constant over a binomial

LESSON 7

AIM/TOPIC(S): More about variables; Simplifying algebraic equations by combining like terms

PREREQUISITE KNOWLEDGE: Review all previous material before beginning this lesson. By this point, the student will hopefully realize that no algebra topic is "in a vacuum." Each topic in some way relates to a previous or future topic.

WARM-UP: Survey the student to get his/her ideas on why a variable is called a variable, and in what way it actually varies. Then present the equation $-3x + 4x + 8x = 8 + 22 - 7$, and ask the student for his/her suggestions on how to proceed with solving for x. Use any responses to lead into the lesson.

MOTIVATION: This is a very short lesson which serves to review and expand upon previous concepts.

POINTS TO ELICIT: Revisit the warm-up matter about variables. Ensure the student understands that any letter can be chosen. If the same letter appears more than once within the same problem, it will take on the same value for each occurrence. If the same letter is used in different problems, it can take on a different value for each. It will only be a coincidence if it takes on the same value in more than one problem.

In an expression, the value of the variable will either be stated in the question, or we are free to choose any value for which we want to evaluate the expression. In an equation, we cannot just choose a value for the variable, nor will such a value actually vary, at least not within the same problem. With an equation, we must determine what value of the variable will make the equation true, meaning the left side equals the right.

Revisit the second warm-up problem. We already practiced combining like terms within an expression. Elicit that we can and should combine like terms on each side of the equals sign before proceeding to solve for x.

Ensure the student understands that we must combine like terms on one side of the equation, and then the other. Later we will learn what to do when an equation has both variable and constant terms on both sides of the equals sign.

DEMONSTRATIVE EXAMPLES / PRACTICE EXERCISES:

If the student does not appear confident about solving problems like the second warm-up problem, make up additional problems which are similar. For now, put only variable terms on one side, and only constants on the other.

At this point in the material, it starts to become very apparent if previous lessons have not been fully mastered. It also becomes apparent if material from the basic math book in this series was not fully mastered, such as how to add -5 + 3.

Take time to ensure that there are no gaps in the student's math knowledge. If there are, fill them before moving ahead. If you don't, at some point in the next few lessons, the student will throw his/her hands up in the air and declare, "I hate this stuff, I don't get it." Do not allow it to get to that point. Ensure that both the student and his/her parents understand that there is no way to avoid learning math step-by-step.

INSTRUCTOR'S NOTES:

LESSON 8

AIM/TOPIC(S): Solving two-step algebraic equations

PREREQUISITE KNOWLEDGE: Many tutors start this lesson even though the student has not mastered one-step equations or basic algebra concepts in general. In particular, many students feel that although 2x means "2 times x," it is impossible to know what 2a means since there is no operation symbol present. Do not begin this lesson if the student is not ready.

WARM-UP: Present the equation $2x + 3 = 11$. Ask the student to mentally solve for x, perhaps using a guess-and-check method. Then elicit his/her ideas on how we could solve the equation algebraically, remembering that we must work to "undo" whatever is being done to the x in an effort to get it alone on one side. Use any response to lead into the lesson.

MOTIVATION: Basic two-step equations are extremely common in algebra. They are very easy, but many students insist on unnecessarily complicating the matter.

POINTS TO ELICIT: Many students will deduce an answer of 4 for the warm-up problem, and protest having to learn any special procedure. As before, explain that we need to have a procedure in case we are faced with more difficult numbers.

Elicit that the warm-up problem cannot be solved in just one step. Two things are being done to x. First, it is being multiplied by 2, and then 3 is being subtracted from that product. While there is more than one way to proceed, the simplest is to "undo" the problem in reverse. First we will subtract 3 from each side. Once that is done, we will divide both sides by 2.

After subtracting 3 from both sides we have $2x = 8$. Some students will protest and say that we should have subtracted 3 from both the 3 as well as the 2x. Explain that if we did that, we would effectively be subtracting 6 on the left. Also, even if we wanted to subtract 3 from 2x, we would have to leave the result as $2x - 3$ since those are unlike terms.

We now have a one-step equation like we've worked with. Divide each side by 2 to get $x = 4$. Check by substituting into the original equation. Both sides are equal so we are correct.

Some students will ask why we can't first divide both sides by 2, and then subtract 3 from each side. Convince the student that we can, but if we do, we have extra complications. First, we cannot just divide the 2x by 2 in a vacuum. We would have to divide the entire left side by 2, and that 2 would have to be distributed over both the 2x and the 3. Remind the student that this is similar to the distributive property of multiplication.

We would end up with $(2x/2) + (3/2) = (11/2)$. The twos would "cancel" leaving x, but we now have fractions to deal with. It is good practice to continue solving the problem from this point. We still get $x = 4$. Convince the student that the first method is just simpler, and less error-prone.

DEMONSTRATIVE EXAMPLES / PRACTICE EXERCISES:

At this level, all two-step equations will follow this format. Just drive home the concepts and procedure. The main book has practice exercises if necessary, but make up your own as well. Include problems which involve subtraction as well as addition, as well as problems involving negatives. It is fine if answers work out to be fractions or negatives.

INSTRUCTOR'S NOTES:

LESSON 9

AIM/TOPIC(S): Equations with variables and constants on both sides

PREREQUISITE KNOWLEDGE: Before starting this lesson it is essential that all previous material have been mastered.

WARM-UP: Present the equation $7x - 8 = 3x + 4$. Ask the student how this equation is different than the ones that we worked with previously. Ask the student to practice what we learned about combining like terms prior to proceeding with solving, and see if s/he realizes that we cannot combine terms "across" the equals sign. We can only combine like terms which are on the same side of the equals sign.

Assess the student's ideas on how we might proceed with solving this equation. Use any response to lead into the lesson.

MOTIVATION: Equations like the one in the warm-up are very common in algebra, but are confusing for many students. Plan on spending extra time on this lesson.

POINTS TO ELICIT: Revisit the warm-up problem. Elicit that we cannot combine like terms "across" the equals sign. There is nothing that can be combined on the left, and nothing that can be combined on the right.

To solve this equation, we must "move" all of the variable terms onto one side, and "move" all of the constants onto the other. It is essential to stress that it doesn't matter which side we choose for which. We will do it both ways to prove this.

Let's decide to get all of the variables onto the left, and the constants on the right. Elicit that we can't just move the 3x over to the left by rewriting it there. We must still follow the algebraic procedure of doing the same thing to both sides of the equation. To "get rid of" the 3x on the right, we can subtract 3x from both sides. That will give us $4x - 8 = 4$. Review the previous lesson in which we discussed that we only subtract 3x from the variable terms, and not the constant terms (which wouldn't be permissible even if we wanted to).

Now we must "move" the 8 over to the right. Seeing that 8 is being subtracted on the left, the inverse is to add 8 to both sides. We get $4x = 12$. Finally, we divide both sides by 4 to get $x = 3$. As always, check via substitution.

For practice, have the student solve the problem such that the variables are accumulated on the right, and the constants on the left. We get $-12 = -4x$. Divide each side by -4 to get $x = 3$, just like before.

Do not allow the student to maintain that the idea that only one of the two demonstrated methods is correct. The only difference is that one will lead to negatives, which is nothing to fear, but of course it is essential it is for the student to have no difficulty at all with signed number arithmetic.

PRACTICE EXERCISES: The main book has practice exercises, but make up your own as needed. Follow the model of the warm-up problem using various values. It is fine if the answer works out to be a terminating or repeating decimal.

INSTRUCTOR'S NOTES:

LESSON 10

AIM/TOPIC(S): Finding an unknown value which will result in a list of values having a specified mean

PREREQUISITE KNOWLEDGE: If it seems the student has forgotten how to compute a average (mean) in general, review the corresponding material from the first book of this series.

WARM-UP: Ask the student to imagine having received test grades of 63, 74, 89, and 94. S/he is going to take a fifth test, and would like for his/her average over the five tests to be 85. Survey the student to get his/her intuition on whether or not it will be possible, and if so, what grade would be necessary on the fifth test. Use any response to lead into the lesson.

MOTIVATION: This lesson teaches how to solve a very popular question which appears on standardized algebra tests.

POINTS TO ELICIT: Revisit the warm-up problem. Walk the student through converting the problem into an algebraic equation. We don't know what the unknown grade is, so let's call it x. We know that to compute the average, we will need to add up all 5 scores, including the unknown. That gives us $63 + 74 + 89 + 94 + x$.

The next step is to divide all of that by the number of scores, which is 5. That gives us the average, and we want that average to be 85. We set up our equation as follows: $\frac{63 + 74 + 89 + 94 + x}{5} = 85$. Combine like terms to get $\frac{320 + x}{5} = 85$. Multiply both sides by 5 to "cancel" the denominator of 5. Elicit that when we do this, we don't distribute the 5 over the

320 + x because it "cancels" before we would do that. We now have 320 + x = 425 which yields x = 105. Hopefully there will be an opportunity for extra credit on the fifth test! Remind the student to check his/her answers for reasonableness.

PRACTICE EXERCISES / DEMONSTRATIVE EXAMPLES: The main book has some practice exercises, but just review the pattern and steps presented in this lesson. At this level, all problems will be of this format. Practice solving problems in which one or more of the given test scores are 0. Many students automatically dismiss such scores saying, "They don't count." Explain that they must be counted, and demonstrate the extent to which they pull down an average.

Also practice some problems involving a very low grade among many high ones. Many students don't realize the extent to which a low grade can really hurt their average, unless it is being averaged along with a large quantity of test scores.

IDEAS FOR EMBELLISHMENT: Time permitting, have the student run some experiments using his/her own test scores, or made-up scores. Ensure that any problems which are set up follow the format of the one in this lesson.

POINTS TO REEMPHASIZE: Remind the student that when we divide the sum of the test scores by the number of scores, the number of scores must include the unknown score.

Do not allow the student to introduce complications such as whether or not the teacher has a policy of dropping the lowest grade. Certainly a problem would specify such, but at this level the problems will all follow the model presented.

INSTRUCTOR'S NOTES:

LESSON 11

AIM/TOPIC(S): Solving algebraic equations involving a fraction times a variable

WARM-UP: Elicit the student's ideas on how to solve for x in the equation $\frac{2}{3}x = 21$. Use any response to lead into the lesson.

MOTIVATION: This simple lesson teaches how to handle equations like the one in the warm-up. They are very easy as long as the student follows one simple rule and doesn't panic.

POINTS TO ELICIT: We know how to solve a problem such as 3x = 21. Just divide both sides by 2. It follows that to solve the warm-up problem we should divide both sides by ⅔. Elicit that to do this, we would have to compute 21 ÷ ⅔, and remind the student that we know how to do that. We get 14. If necessary, review the steps for dividing by a fraction.

While the above method works, there is a method that is quicker. Instead of dividing both sides by 2/3, we can multiply both sides by 3/2, the reciprocal of 2/3. Elicit that when we do this, the left side becomes 1x, or just x. On the right we get 14 just like we got with the other method. Ensure that the student has no difficulty at all doing the actual arithmetic of multiplying an integer times a fraction. If s/he does, review the material from the basic math book in this series as needed.

PRACTICE EXERCISES: The main book has practice exercises, but focus on the concept. In problems like this, just multiply both sides by the reciprocal of the variable's coefficient.

INSTRUCTOR'S NOTES:

LESSON 12

AIM/TOPIC(S): The simple interest formula: $I = Prt$

WARM-UP: Survey the student to determine his/her experience with general concepts of banking (principal, interest, APY/APR, etc). Use any responses to lead into the lesson.

MOTIVATION: This quick lesson covers a popular topic which will also serve to review some basic algebra.

POINTS TO ELICIT: Explain the general concept of how invested money earns interest. In this lesson we will learn what is called simple interest formula. Do not allow the student to worry about compound interest, or to in any way make this easy lesson more complicated than necessary.

Present the formula $I = Prt$. P stands for principal. Define it as the amount of money that we will invest for some time. I is the interest. It is the amount of money that the bank (or investment company) will give us after investing the money for a period of time. It does not include the principal.

Define r as the (annual) interest rate (APR/APY) offered by the investment. If it is given as a percent, we should convert it to a decimal to facilitate calculations. Review the basic math book in this series if the student needs help with that step.

Define t as the amount of time (in years) that the principal will be invested. The formula is based on the fact that t will be given in years. If the money is to be invested for 6 months, t will be 0.5. At this level, t will either be a whole number of years, or something simple like 0.25 to represent 3 months.

In basic problems involving the simple interest formula, we will be given three out of the four pieces of information. All we need to do is substitute the given values into the formula, and then solve for the missing value. We must remember to convert r into a decimal, and t into years if given in months.

DEMONSTRATIVE EXAMPLES / PRACTICE EXERCISES: Make up some examples, referring to the main book if needed. If we are given P, r, and t, we just multiply them to get I. If we are given I and two of the three other variables, we just multiply on the right, then divide to get the missing variable.

Always check the answer for reasonableness. Be certain to answer the specific the question asked. For example, we may be asked to compute just the interest earned, or how much money we will have in total at the end of the investment. In the latter, we must remember to add the principal back in.

IDEAS FOR EMBELLISHMENT: If time permits, the student may enjoy running "real-world" scenarios based on current interest rates for different types of investments. Many students don't realize how slowly their money grows in an investment with a low interest rate. Consider obtaining the permission of the student's parents before doing anything which could be construed as giving financial advice or passing judgment.

POINTS TO REEMPHASIZE: The variable t is always given in years. It is common for a question to specify a time duration of 6 months to trick the student into thinking that $t = 6$. Ensure that the student knows how to convert a percent into a decimal, and will do so. For example, the formula won't work if 4% is not converted to 0.04, and if the student just uses $r = 4$.

INSTRUCTOR'S NOTES:

LESSON 13

AIM/TOPIC(S): Algebra problems involving proportions

WARM-UP: Present the proportion 2/5 = 6/15. Ask the student how we can be certain that the two fractions are in fact equal. Then present the proportion 2/5 = x/17. Ask the student why it is difficult to compute the answer in one's head. Use any responses to lead into the lesson.

MOTIVATION: Proportion problems are common in algebra. They are easy as long as one simple concept is remembered.

POINTS TO ELICIT: Revisit the first warm-up problem. Review the concept that in a proportion, the cross-products are equal. If the student expresses any doubt or confusion, review all of the related material from the first book until it is clear.

Revisit the second problem. Elicit that it is not as simple as saying that the denominator on the right is a certain number of times bigger than the denominator on the left. We can solve the proportion by setting up an equation in which the cross-products are equated. We get 5x = 34. Divide each side by 5 to get x = 6.8. Always check answers for reasonableness.

DEMONSTRATIVE EXAMPLES / PRACTICE EXERCISES: The main book has some practice exercises, but just drive home the concept and procedure. All proportion problems at this level will follow this format. Ensure that the student will not be flustered if an answer works out to be a terminating or repeating decimal.

INSTRUCTOR'S NOTES:

LESSON 14

AIM/TOPIC(S): Using the distributive property to multiply a constant times a binomial involving a variable

PREREQUISITE KNOWLEDGE: It is essential that the student have no difficulty at all with signed number arithmetic.

WARM-UP: Ask the student to simplify (evaluate) the expression 3(2 + 5) using the distributive property of multiplication over addition. Use any response to lead into the lesson.

MOTIVATION: This lesson explains why we sometimes need to use the distributive property. When presented in the first book, it was difficult to rationalize since no variables were involved. The concepts presented in this lesson will be used repeatedly in algebra, so ensure that the student grasps them.

POINTS TO ELICIT: Revisit the warm-up problem. The 3 can be distributed over the 2 and 5, giving us 3(2) + 3(5) which is 21. We would get the same answer if we first added 2 + 5, but that only works because they are combinable like terms.

Present the expression 3(x + 2). Elicit that we cannot combine the x and the 2 even if we wanted to. They are unlike terms. Ensure that the student is not harboring the idea that we can combine them to get 2x. Explain that if this expression was part of an algebraic equation, one of our first tasks would be to eliminate the parentheses. This can be a bit of a hard sell, and admittedly many equations allow for alternate means of solving. Just try to get the student to accept that when s/he sees an expression like 3(x + 2) in any context, we should use the distributive property to remove the parentheses.

Following the steps presented in the first book, we can distribute to get $3x + 3(2)$, which further simplifies to $3x + 6$. If the expression had been part of an equation, perhaps being set equal to a constant, we could then proceed with the two-step equation like we learned in a previous lesson.

The biggest challenge with this topic is when we have to apply the distributive property to an expression which involves negative numbers and subtraction. The subtraction part itself is not really the issue, since all we do is use a subtraction sign in place of plus sign when we distribute. The issue is how to correctly handle any negatives. This is a major difficulty for most students, and extra time should be spent on it.

Let's look at the expression $-6(x - 5)$. Let's break it down at the most detailed level. First, the -6 must multiply the x, giving us $-6x$. Then, the -6 must multiply the 5, giving us -30. Note that since we are treating the dash between the x and the 5 as a subtraction operation, we are treating the 5 as positive. We could have instead converted $x - 5$ into $x + (-5)$. That concept was covered at length previously. We obtained products of $-6x$ and -30 which must be connected via subtraction, as was the original expression. We get $-6x - (-30)$ which would usually be rewritten as $-6x + 30$.

DEMONSTRATIVE EXAMPLES / PRACTICE EXERCISES:
The main book offers practice exercises, and it is not hard to make up your own. However, there is no point to endlessly drilling and having the student guess at whether an answer involves positives or negatives. Spend as much time as necessary reviewing the concepts of this lesson, as well as the material from the first book on signed number arithmetic.

INSTRUCTOR'S NOTES:

CHAPTER FOUR

Lesson Plans for "Exponents in Algebra"

Lessons 15 to 19

Topics Covered in This Chapter:

Review of exponents; Problems of the form $x^a \cdot x^b$, x^a/x^b, and $(x^a)^b$; Simplifying x^{-a}; Why $x^0 = 1$; Trick problems of the form $x^a + x^b$; Multiplying terms of coefficients and different variables; Distributing a variable with coefficient over a binomial

LESSON 15

AIM/TOPIC(S): Review of exponents; Problems of the form $x^a \cdot x^b$ and $(x^a)^b$; Trick problems of the form $x^a + x^b$

PREREQUISITE KNOWLEDGE: It is essential that the student at least mostly remembers the material from the first book on exponents. This lesson works with exponents more abstractly.

WARM-UP: Survey the student to determine his/her understanding of the concept of exponents. Use any response to lead into the lesson.

MOTIVATION: Exponents are a huge component of algebra. Do not allow the student to dismiss this lesson as though it is an isolated topic.

POINTS TO ELICIT: Based on the warm-up, review related material from the first book as needed. Present the expression $x^2 \cdot x^3$, and elicit the student's ideas on how we can simplify it. Lead the student toward thinking about what each exponential term actually means. The first one means x · x. The second means x · x · x. Those two expressions must get multiplied together, giving us x · x · x · x · x. We can use an exponent to condense that to x^5.

Elicit that $x^a \cdot x^b = x^{(a+b)}$. We add the exponents because that represents how many copies of the base we end up with. Most students feel as though they must memorize the formula. Stress that it is not really necessary to do so if it is understood. Also remind the student that the x could be an actual value, for example, $7^2 \cdot 7^3 = 7^5$. The answer would be left in that form.

Present the expression $(x^2)^3$. Elicit the student's ideas on how to simplify it. Many students will just guess x^5. As before, lead the student to think about what the expression actually means. We know that x^2 means x · x. The cubing tells us to take the x · x and multiply it times itself for a total of three copies. We get x · x · x · x · x· x which is x^6. Elicit that $(x^a)^b = x^{ab}$. We multiply the exponents because that represents how many copies of the base we get. Again, instead of memorizing the formula, it is better if the student understands it.

Present the expression x^a+x^b, and elicit the student's ideas on how to simplify it. Answers will vary, but will mostly likely be incorrect. Make the problem more concrete by presenting the example 2^4+2^5, and having the student simplify it to 16 + 32 which is 48. Elicit that there is no way we can get 48 by raising 2 to a given power. Drive home the concept that any problem of the form x^a+x^b cannot be simplified, in the sense that it cannot be condensed into a base raised to a power.

In the above scenario, if the base is a variable whose value is not known, we have no choice but to leave the expression as is. If the base is a constant, we could certainly evaluate the expression by raising each term to the given powers and adding, but we will then have no way of rewriting that answer using an exponent. Any problem of the form x^a+x^b is essentially an attempt to trick the student into thinking that it can be simplified to $x^{(a+b)}$, or x^{ab}, or something similar.

DEMONSTRATIVE EXAMPLES / PRACTICE EXERCISES:
The main book has some practice exercises, but as always, just work to ensure that the student has internalized the concepts.

INSTRUCTOR'S NOTES:

LESSON 16

AIM/TOPIC(S): Problems of the form x^a/x^b

WARM-UP: Review the previous lesson, and then survey the student to get his/her ideas on how we would simplify x^5/x^3. Use any response to lead into the lesson.

MOTIVATION: This lesson builds upon the previous one. It is essential for the student to fully master it.

POINTS TO ELICIT: Elicit that when we multiplied exponential terms with common bases, we added the exponents. It makes sense, then, that when we divide terms with common bases, we should subtract the exponents. Let's see why.

Rewrite the warm-up problem using a fraction bar. Elicit that we can "cancel" pairs of x's in the numerator and denominator, but that what is really happening is that each pair is x/x which is 1, and need not be written. After "canceling" three times, we are left with x · x in the numerator, and an implied 1 in the denominator which doesn't matter.

We are really just left with x^2. The exponent of 2 was simply the difference between 5 and 3. We could have subtracted the exponents to determine how many more x's we had in the numerator than the denominator. Elicit that $x^a/x^b = x^{(a-b)}$. In the next lesson we will handle the case of b > a.

PRACTICE EXERCISES: There is little to practice with this topic. Rather than memorize the formula, the student should strive to understand why it works.

INSTRUCTOR'S NOTES:

LESSON 17

AIM/TOPIC(S): Simplifying x^{-a}; Why $x^0 = 1$

PREREQUISITE KNOWLEDGE: It is essential that the student have mastered the material on exponents presented thus far.

WARM-UP: Review the previous lesson, and then survey the student to get his/her ideas on how we would simplify x^3/x^5. Use any response to lead into the lesson.

MOTIVATION: This lesson builds upon the previous one, and teaches important concepts which are common in algebra.

POINTS TO ELICIT: Rewrite the warm-up problem using a fraction bar. Elicit that we can "cancel" pairs of x's in the numerator and denominator, but what is really happening is that each pair is x/x which is 1, and need not be written.

After "canceling" three times, we are left with x · x in the denominator, and an implied 1 in the numerator which we must write because we can't have an empty numerator. This is different than when we had 1 in the denominator which we didn't have to write since anything divided by 1 is itself.

We are really just left with $1/x^2$. Recall our formula of $x^a/x^b = x^{(a-b)}$. In this case we had x^3/x^5 which would give us x^{-2}. Elicit why this makes sense. In this problem, the x's in the denominator exceeded the x's in the numerator by 2. Review that when we simplified this problem by "canceling" pairs of x's, we ended up with $1/x^2$. This shows that x^{-2} and $1/x^2$ are equivalent.

Elicit the formula $x^{-a} = 1/x^{a}$. Whenever we end up with a negative exponent, it tells us the extent to which the x's in the denominator exceed the x's in the numerator. Since the entire numerator will be "wiped out" after "canceling" (leaving us with implied 1's), we put a 1 in the numerator. We can then make the exponent positive and put it in the denominator to show how many x's remained after "canceling."

DEMONSTRATIVE EXAMPLES / PRACTICE EXERCISES:
There is little to demonstrate or practice once the concept has been learned. Try to convince the student that when s/he ends up with a negative exponent, it is usually expected that the term will be placed under a numerator of 1, with the exponent made positive. Until later math, the reason may not make much sense.

In the first book, it was stated without explanation that $x^0 = 1$. Let's demonstrate this informally by example. Ask the student to simplify x^3/x^3 by "canceling" matching pairs. All the x's will cancel, leaving us with 1/1 or 1. Now have the student simplify the problem using the formula which tells us to subtract the exponents. We get x^0. From this we can infer that x^0 must equal 1. Allow the student some time to digest this.

POINTS TO REEMPHASIZE: This short lesson allows for additional review time. Use that time to review all of the previous material to this point. Again, it is not sufficient for the student to have only a vague idea of what is going on. Either the student fully internalizes all of the concepts or s/he doesn't. The latter will lead to failure and frustration, so do not allow things to get to that point.

INSTRUCTOR'S NOTES:

LESSON 18

AIM/TOPIC(S): Multiplying terms which involve coefficients and different variables

WARM-UP: Elicit suggestions for simplifying the expression $8xy^2 \cdot 5x^4y^6$. Use any response to lead into the lesson.

MOTIVATION: This lesson reviews previous material and applies it to a common task in algebra.

POINTS TO ELICIT: Review that multiplication is commutative. Revisit the warm-up problem, eliciting that it is just a string of items to be multiplied. A middle dot and/or the absence of an operation symbol implies multiplication.

We can multiply the components of the expression in any convenient order. With that in mind, it makes sense to rearrange the problem (perhaps mentally) into $(8 \cdot 5) \cdot (x \cdot x^4) \cdot$ $(y2 \cdot y6)$. We already know how to handle that, but review as needed if the student does not. We get $40x^5y^8$. Notice how the x and y terms are handled separately since they are unlike.

DEMONSTRATIVE EXAMPLES / PRACTICE EXERCISES:
The main book has additional examples. Difficulties with this lesson are usually the result of not having mastered the previous material. For example, the student must know how to multiply exponential terms with common bases, and that a variable by itself has an implied exponent of 1. S/he must understand the commutative property of multiplication, and that this problem in no way involves the distributive property.

INSTRUCTOR'S NOTES:

LESSON 19

AIM/TOPIC(S): Using the distributive property to multiply a term involving a coefficient and variable times a binomial

WARM-UP: Review the previous lessons on the distributive property. Ask the student to try applying the property to the expression 4x(6x + 7), and then to do the same for the expression -3x(5x – 6). Use any responses to lead into the lesson.

MOTIVATION: This lesson reviews important previous concepts and applies them to a task which is very common in algebra. Do not allow the student to treat the lesson as trivial.

POINTS TO ELICIT: Revisit the first warm-up problem. The 4x must multiply the 6x and the 7, with the two resulting products connected with addition. We must compute (4x · 6x) + (4x · 7). We get $24x^2$ + 28x. Do not proceed if the arithmetic or basic algebra is troublesome. We must leave our answer as is since the two terms are unlike and cannot be combined.

Revisit the second warm-up problem. The -3x must multiply the 5x and the 6, with the two resulting products connected with subtraction. We must compute (-3x · 5x) – (-3x · 6). We get $-15x^2$ – (-18x), typically rewritten as $-15x^2$ + 18x. Do not proceed if the signed number arithmetic is troublesome.

DEMONSTRATIVE EXAMPLES / PRACTICE EXERCISES:
The main book has additional examples. Difficulties with this lesson are usually the result of not having mastered the previous material, including signed number arithmetic.

INSTRUCTOR'S NOTES:

CHAPTER FIVE

Lesson Plans for "Solving and Graphing Algebraic Inequalities"

Lessons 20 to 21

Topics Covered in This Chapter:

Introduction to inequalities; Solving algebraic inequalities; When to "flip" the direction of the inequality; Graphing inequalities on a number line

LESSON 20

AIM/TOPIC(S): Introduction to inequalities; Solving algebraic inequalities; When to "flip" the direction of the inequality

WARM-UP: Present the inequality $2x > 8$. Explain that it could be thought of as a special type of equation. Without solving algebraically, ask the student what value(s) of x will satisfy the inequality. Use any response to lead into the lesson.

MOTIVATION: Inequalities are common in algebra. They are easy to handle, as long as a few simple concepts are mastered.

DEMONSTRATIVE EXAMPLES / POINTS TO ELICIT:
Revisit the warm-up problem. Try to elicit that any value of x greater than 4 will satisfy the inequality. The value 4 itself is excluded because when x is 4, the left side is equal to but not greater than 8. Most students are OK with this, with the difficulty only setting in when negative numbers are involved.

Introduce the $\geq$ and $\leq$ symbols. Some students don't like the idea of the "or equal to" component and ask, "Well which is it?' Elicit that had the previous inequality been $2x \geq 8$, the solution would have been $x \geq 4$. Any value of x which is greater than 4 would work, and now 4 itself would work since it is fine if the left side equals the right side. Elicit that any value of x which is less than 4, including negative values, would not work in the previous in equality.

Demonstrate that inequalities are solved algebraically the same way we solve equations, with an important exception described after this example. Solve: $4x + 7 < 27$. Subtract 7 from each side, giving us $4x < 20$. Divide each side by 4 to get $x < 5$.

We solved the inequality like an equation. As the last step, we put back the less-than sign. For practice, have the student substitute various values of x until convinced that all values less than 5, including 0 and negatives, satisfy the inequality.

Here is the important exception. When solving an inequality, if we multiply or divide by a negative value, we must flip the direction of the inequality sign, for example, $\geq$ becomes $\leq$. We only do this when we multiply or divide by a negative as part of the solving process. We do not do this if the inequality just happens to involve a negative. For example, the solution to $3x > -18$ is $x > -6$. We divide each side by 3, and leave the inequality sign as is. For practice, have the student substitute values of x which are greater and less than -6 until convinced that we were correct in leaving the inequality sign as is.

Here is an example where we do have to flip the inequality sign. Solve $-2x \leq 6$. Divide each side by -2 to get $x \leq 3$, but since we divided by a negative, we must flip the inequality sign to get $x \geq 3$. Most students ask why this works, or why we must flip the sign. It is usually best to have the student test out values of x which satisfy and dissatisfy the solution until convinced that the sign needed to be flipped. If the student insists on a better explanation for flipping the direction of the inequality sign, you could informally explore how division or multiplication by a negative essentially reverses things due to the mirror-image nature of the number line.

PRACTICE EXERCISES: The main book offers some practice exercises, but all that is essential is for the student to fully grasp and retain the concepts of this lesson.

INSTRUCTOR'S NOTES:

LESSON 21

AIM/TOPIC(S): Graphing inequalities on a number line

PREREQUISITE KNOWLEDGE: Review the previous lesson.

WARM-UP: Illustrate the representation of $x < 2$ on a number line via an open circle and left-pointing arrow. Refer to the main book if necessary. Then do this for $x \geq -3$ via a closed circle and right-pointing arrow. Most students "get" this topic with little or no difficulty, so proceed into the lesson.

MOTIVATION: This quick and easy lesson teaches how to graph inequalities on a number line. Most students enjoy it.

POINTS TO ELICIT / DEMONSTRATIVE EXAMPLES:
Revisit the warm-up. Elicit that we use a closed circle to represent "or equal to," and an open circle to represent inequalities which don't include that option. Most students find this intuitive. Elicit that the circle is placed at the designed value, and we use a thick left or right-pointing arrow to represent less than or greater than, respectively. If the student is concerned about the nuances of drawing the arrow, just explain that the test-grader needs to clearly see what is being represented, but that there are no specific guidelines.

Illustrate how to represent $-1 < x \leq 2$ via an open circle at -1, a closed circle at 2, and a thick line connecting the two. Again, students typically have no trouble with this.

PRACTICE EXERCISES: The main book has practice exercises, but extra time is better spent reviewing previous topics.

INSTRUCTOR'S NOTES:

CHAPTER SIX

Lesson Plans for "FOILing and Factoring"

Lessons 22 to 27

Topics Covered in This Chapter:

Algebra definitions; Multiplying binomials with the FOIL method; Factoring trinomials via "reverse FOIL"; Factoring the difference of two squares; "Pulling out" common factors; Factoring the square of a binomial; Simplifying algebraic fractions

LESSON 22

AIM/TOPIC(S): The FOIL method and related definitions

WARM-UP: Present the expression $(x + 3)(x + 4)$. Survey the student for his/her ideas on how we can use the distributive property to remove the parentheses, reminding him/her that we must keep track of all combinations which need to be multiplied. Use any response to lead into the lesson.

MOTIVATION: This lesson introduces terminology and a new procedure which is used constantly in algebra. Ensure that the student understands the importance of this lesson.

DEFINITIONS: Essential terminology must be learned before starting this lesson. A monomial is a one-term expression. A term can be a constant, a variable, or several constants or variables connected via multiplication or division. Elicit some examples, referring to the main book as needed.

A binomial is an expression of two terms connected by addition or subtraction. A trinomial is an expression of three terms connected by addition and/or subtraction. A polynomial describes any multi-term expression. Elicit examples of each.

POINTS TO ELICIT: Revisit the warm-up expression. Elicit that had we been dealing with $3(x + 4)$, we would have used the distributive property like we learned. In the warm-up, though, we are dealing with a binomial times a binomial. Try to convince the student that in addition to the 3 multiplying each term of the second binomial, the x of the first binominal must get multiplied by each term of the second binomial. Elicit that we have four products to keep track of.

Introduce the idea of the FOIL method for handling the warm-up problem. Explain that the acronym tells us to multiply the first terms (yielding x^2), the outer terms (yielding 4x), the inner terms (yielding 3x), and the last terms (yielding 12). Elicit that we have covered all of the possible combinations.

Explain that the final step of the FOIL method is to add the four products which we obtained. Stress that FOIL represents the sum of products. We get $x^2 + 4x + 3x + 12$, and after combining like terms we get $x^2 + 7x + 12$. Explain that we often informally use FOIL as a verb. We started with a product of two binomials, and we FOILed them to get a trinomial.

Many students will simply ask, "What for?" As before, explain that it ties into later math, and that in the next lesson our task will be to follow this procedure in reverse, (i.e., start with a trinomial, and "reverse-FOIL" it to get a product of binomials).

DEMONSTRATIVE EXAMPLES: For additional practice, have the student use FOIL on (x + 5)(x – 8). S/he must be very careful with the signed number arithmetic. To avoid confusion, consider starting by converting x – 8 into the equivalent x + (-8) so that each binomial involves addition. We get $x^2 + (-8)x + 5x + (-40)$. After combining and rewriting in the more common format, we get $x^2 - 3x - 40$.

PRACTICE EXERCISES: The main book has additional practice exercises. The key point to reemphasize is that FOIL is a sum of products. It's really F+O+I+L where each letter represents a product. Most students don't have trouble with FOIL itself, but just get very flustered with problems involving negatives. As always, review previous material as needed.

INSTRUCTOR'S NOTES:

LESSON 23

AIM/TOPIC(S): Factoring polynomials using "reverse-FOIL"

PREREQUISITE KNOWLEDGE: Ensure that the previous lesson is completely mastered before starting this one.

WARM-UP: Revisit the warm-up problem from the previous lesson which led to $x^2 + 7x + 12$. Survey the student for his/her ideas on how we could use a reverse-FOIL technique to work backward to get from that trinomial to the original product of binomials. Use any response to lead into the lesson.

MOTIVATION: This lesson teaches how to perform one of the most common tasks in algebra—factoring trinomials.

POINTS TO ELICIT: Revisit this lesson's warm-up task. What we are being asked to do is called "factoring the trinomial." We must "break it up" into a product of binomials (i.e., its factors). Many students believe that there is some secret formula for performing this task, but there truly isn't. The matter involves literally working backwards, and logically deducing what the original pair of binomials must have been.

At first the student will use a guess-and-check method for every possible combination which might be correct, but with practice, the correct choice will become obvious more quickly. It is a matter of building one's math "intuition," as well as becoming adapt at mental arithmetic.

In our example, we can start by setting up our product of binomials as (x   )(x   ). That is only way we can get x^2.

We can then think about what the last terms could be in order for us to end up with +12. They could be 1 and 12, 2 and 6, or 4 and 3. They could both be positive, or they could both be negative since a negative times a negative yields a positive.

At this point, the student is advised to test out each combination while using common sense to guide him/her to the next combination to try. If we use +1 and +12, we will not be able to end up with the 7x that we need. We would end up with a middle term of +13x. Making both negative won't help. This same logic can be applied to 2 and 6. The only combination which will work is +4 and +3. If both were negative we would end up with +12 which we want, but -7x which we do not. We say that x^2 + 7x + 12 can be factored as (x + 3)(x + 4). The importance of this task is explained in a later lesson.

DEMONSTRATIVE EXAMPLES / PRACTICE EXERCISES:
For practice, have the student factor x^2 + x – 12. The student will have to be extra careful with handling the negative. Elicit that the last two terms will have to be of opposite signs since that is the only way we can end up with a negative. This requires the additional task of determining which number should be negative, and which should be positive. It will make a difference when adding the inner and the outer products. Have the student keep experimenting until s/he determines that the only proper factorization is (x – 3)(x + 4). Note that it doesn't matter if the second binomial was written first.

POINTS TO REEMPHASIZE: In later math, the student will be required to factor more challenging polynomials such as ones which begin with ax^2 where a > 1. For now, though, all polynomials will follow the format of those in this lesson.

INSTRUCTOR'S NOTES:

LESSON 24

AIM/TOPIC(S): Factoring the difference of two squares

PREREQUISITE KNOWLEDGE: Review the previous lesson.

WARM-UP: Present the binomial $x^2 - 49$. Elicit that this problem doesn't involve an x term like the ones in the previous lesson, but that it would not cause any harm if we inserted an implied +0x as the middle term. With that in mind, have the student try using reverse-FOIL to factor what is now a trinomial. Use any response to lead into the lesson.

MOTIVATION: This short lesson teaches how to factor binomials which are the difference of two squares.

POINTS TO ELICIT: Revisit the warm-up. Elicit that we are dealing with the difference of two squares. x^2 is the square of x, and 49 is the square of 7. How can use we reverse-FOIL to factor $x^2 + 0x - 49$ (which includes the implied +0x)? Again, we must experiment until we develop our math intuition. We can start with (x   )(x   ). We know that the last two terms must have opposite signs so that we can end up with a negative. The only choices for values are 1 and 49, and 7 and 7. Experimenting will quickly prove that the only combination which will work is (x – 7)(x + 7).

DEMONSTRATIVE EXAMPLES / PRACTICE EXERCISES:
For now, all problems of this type will follow this format. For example, $x^2 - 81$ factors to (x – 9)(x + 9). We used values with opposite signs, each of which was the square root of the value present in the original binomial.

INSTRUCTOR'S NOTES:

LESSON 25

AIM/TOPIC(S): Factoring ("pulling out") common factors

PREREQUISITE KNOWLEDGE: Review the procedure for computing the GCF of two numeric values. Do not start this lesson until the student is comfortable with that task.

WARM-UP: Ask the student to compute the GCF of 5 and 7. Then survey the student for his/her ideas on the GCF of x^5 and x^7. Use any response to lead into the lesson.

MOTIVATION: This lesson extends the concept of GCF to variables and exponential terms. It is very important, and can be tricky. Plan on spending extra time on this three-page lesson and not rushing through it.

POINTS TO ELICIT: Revisit the first warm-up question. The student must have no trouble identifying the GCF as 1. If the student says, "There is none," or "35," or "Is that the same as the LCM?," stop and review the prerequisite material.

Revisit the second warm-up question. Let's think of x as a positive integer greater than 1, but we don't actually have to know what x is. Have the student think about what exponential x-terms will divide into both x^5 and x^7. Let's start with just x itself. Dividing each of those terms by x gives us x^4 and x^6.

This proves that x is a factor of x^5 and x^7. It divided into both of them. Stated another way, we showed that x and x^4 are factors of x^5, and x and x^6 are factors of x^7. Allow the student to sit with this for a while. It is like saying that 3 is a factor of 24, and 3 and 8 are factors of 24.

Let's try a higher potential factor of x^5 and x^7. What happens if we divide each of those terms by x^2? Certainly 2 does not divide into 5 or 7, but that is not the question here. Can we divide x^5 by x^2? We can, and the result is x^3. We can also divide x^7 by x^2. Again, let the student sit with this for a while.

If we continue this procedure, we will see that the GCF of x^5 and x^7 is x^5. If we divide x^5 by x^5 we get 1. If we divide x^7 by x^5 we get x^2. Elicit that we can't divide those terms by any power higher than 5 since it will not divide into x^5.

Elicit that this concept applies regardless of what exponents we are dealing with. For example, the GCF of x^{48} and x^{52} is x^{48}. This has nothing to do with the GCF of the values 48 and 52 since in our problem those values are serving as exponents.

Let's now apply this knowledge to some factoring problems. Whenever we are given a binomial to work with in any context, it is usually advised to look for common factors which can be factored out ("pulled out.") For example, look at the binomial $15x^6 + 20x^8$. The GCF of the coefficients is 5, so we can "pull that out." The GCF of x^6 and x^8 is x^6 so we can pull that out as well.

We have "pulled out" $5x^6$, and must now determine what that term must get multiplied by in order to get back our original binomial. We are effectively using the distributive property in reverse. Using our knowledge of already-covered concepts, we can compute that the original binomial is factored as $5x^6(3 + 4x^2)$. It can be checked with the distributive property.

DEMONSTRATIVE EXAMPLES / PRACTICE EXERCISES: Let's look at another common example. Factor $4x^9 - 12x^{14}$. Following the previous example, elicit that we can factor out $4x^9$ as the GCF of both of those terms. Allow the student to sit with that fact until s/he is convinced.

What must we multiply $4x^9$ by in order to get back our original binomial? To get the first term, we must multiply $4x^9$ by 1. Many students think that we multiply it by "nothing," since we "pulled out" exactly what we need to get back to, but remember that we are dealing with multiplication. $4x^9$ must be multiplied by 1 to get back to $4x^9$. To get back to $12x^{14}$, we must multiply $4x^9$ by $3x^5$. We get $4x^9(1 - 3x^5)$.

The main book has more examples, but as always, it is a matter of mastering the concepts. At this level, the problems will all follow this model. Many students ask why the factored form is preferable to the unfactored. Unfortunately, the answer to that question will mostly not come until later math.

POINTS TO REEMPHASIZE: Even after spending time on this lesson, many students will not understand the difference between 5 and 7, and x^5 and x^7 when it comes to matters of GCF. Continue to stress the concepts presented in this lesson. As always, ensure that any confusion is not just the result of never having learned any or all of the prerequisite concepts.

This is the hardest lesson in the book for most students. We have a bit algebra to go, and then we will start geometry which most students find easier and more interesting. With that said, do not allow the student to think that s/he will be "done" with algebra. Many geometry problems are solved using algebra.

INSTRUCTOR'S NOTES:

LESSON 26

AIM/TOPIC(S): Factoring the square of a binomial; How factoring helps us solve certain problems

WARM-UP: Present the polynomial $x^2 - 10x + 25$. Ask the student to factor it and simplify his/her answer if possible.

MOTIVATION: This quick lesson is a review of previous concepts, with some new notation presented.

POINTS TO ELICIT: Revisit the warm-up. If the previous lesson on reverse-FOIL was mastered, the student should be able to factor the polynomial as (x –5)(x – 5). Elicit that we can represent this more concisely as $(x - 5)^2$.

DEMONSTRATIVE EXAMPLES / PRACTICE EXERCISES:
For additional practice, have the student factor $x^2 + 18x + 81$. The solution is most concisely represented as $\mathbf{(x + 9)^2}$. The main book has additional exercises which all follow this format. Ensure that the student can use FOIL to expand the concise factorization to get back to the original trinomial.

In the next lesson we use factoring in more elaborate problems, but here is an example which the student can think about. Ask him/her to mentally solve the equation $x^2 + 2x - 24 = 0$. It is not easy. Now factor the left side to get (x + 6)(x – 4) = 0. What value(s) of x will make the left side equal 0? Elicit that when x is either -6 or 4, the left side will equal 0, by virtue of the property of multiplication that anything times 0 is 0. The student will work with this more in later math.

INSTRUCTOR'S NOTES:

LESSON 27

AIM/TOPIC(S): Simplifying algebraic fractions

WARM-UP: Present the quotient of polynomials at left, and survey the student for his/her ideas on how to simplify it. If you want to try to "trick" the student, ask him/her to think about what can "cancel."

$$\frac{x^2 + 4x - 21}{x^2 - 8x + 15}$$

MOTIVATION: This lesson involves a task which usually isn't presented until later math. It is covered here for review purposes, and to demonstrate an application of factoring.

POINTS TO ELICIT: Revisit the warm-up. Most students suggest "canceling" the x^2 terms. Review that we can't do that. Even if the denominator only contained x^2, we would still have to distribute it over each term in the numerator. We can't just "cancel" the x^2 terms because they are part of polynomials comprised of addition and subtraction. We can't do any "canceling" until we modify the problem to only involve products. This is one of the benefits of factoring. When we factor, we break up a polynomial into a product of factors.

In this problem we can factor the numerator and denominator to end up with (x – 3)(x + 7) / (x – 5)(x –3). Now that we are left with nothing but products, we can "cancel" the (x –3) terms which appear in both the numerator and denominator. We are left with (x + 7) / (x – 5) which is as far as we go.

DEMONSTRATIVE EXAMPLES / PRACTICE EXERCISES: Time is best spent reviewing all of the material covered so far. We are about to start geometry which involves algebra.

INSTRUCTOR'S NOTES:

CHAPTER SEVEN

Lesson Plans for "Introducing Geometry: Lines and Angles"

Lessons 28 to 31

Topics Covered in This Chapter:

Introduction to geometry; Basic geometry definitions; Types of angles; Complementary and supplementary angles; Parallel and perpendicular lines; Problems involving parallel lines and a transversal

LESSON 28

AIM/TOPIC(S): Intro to geometry; Basic definitions

CONNECTIONS TO LATER MATERIAL:
This lesson forms the basis of the geometry which the student will later study. Ensure that the student has fully mastered it.

WARM-UP / MOTIVATION:
Survey the student to get his/her idea on what it means for something to be two or three dimensional. Some students think a TV screen is 3-D because it represents objects which they know to be such. If the student seems to have at least an intuitive understanding of 2-D vs. 3-D, see if s/he can deduce the number of dimensions in a line and in a point.

POINTS TO ELICIT / DEFINITIONS: Elicit the concept that a rectangle has two dimensions, usually called length and width, and that any flat shape is considered to be 2-D. Many students have a hard time accepting that an elaborate polygon of many sides is still 2-D, so try your best to drive the point home.

Introduce the concept of three dimensions by asking the student to imagine taking a 2-D shape and extending it into the third dimension, perhaps by lifting it up off the table, or outward from the chalkboard. It is ideal if you have some props such as a box, a cube, and a cylinder (e.g., an empty toilet paper roll). Elicit that we can think of the third dimension as height or depth, but it doesn't matter what we call it.

The student may ask why we can't just measure a box along its diagonal, and let that be one of the dimensions. This is even more confusing if s/he later encounters a diagram of a box in which a diagonal measure serves as decoy data. Try to elicit the concept that dimensions are always perpendicular to each other. The term is elaborated on later.

Elicit the concept that if a flat shape has two dimensions, a line must have one dimension, usually called length. Some students will note that a line drawn on paper or the chalkboard certainly has some thickness, so define a line as being infinitely thin, and therefore having no width. Many students have trouble with this abstract concept. We just have to draw lines sufficiently think so that we can see where they are.

Elicit that if a line has one dimension, a point must have zero dimensions. It is just a location in a 2-D or 3-D space. We define a point as having no length or width or any measureable dimension, even though it can be argued that on paper it has a measureable length and width.

Elicit that if we start with a 0-D point, we can extend it in a direction to form a 1-D line, We can then "pull" that line perpendicular to itself to form a 2-D figure. We can then extend that figure perpendicular to itself to form a 3-D solid.

Define a line segment as a line which has a specific endpoint on each end. Define a ray as a line which has a specific endpoint on one end, but extends infinitely forever in the other direction. Show how this is represented with an arrow. Define a line as characterized by extending infinitely in both directions, represented with arrows. Define a plane as a 2-D flat surface such as a piece of paper or a chalkboard.

INSTRUCTOR'S NOTES:

LESSON 29

AIM/TOPIC(S): Using lines to form angles; Angle measurement; Types of angles

CONNECTIONS TO EARLIER/LATER MATERIAL:
This lesson involves some of the terminology from the previous lesson. It also forms the foundation of more advanced work with geometry which will follow.

WARM-UP: Survey the student to get his/her idea of what an angle is, and where s/he encounters angles in everyday life. Ask the student what type of angle s/he encounters the most often, trying to lead him/her to the concept of a right angle. Use any responses to lead into the lesson.

MOTIVATION: Explain that the study of angles is a huge component of geometry. Most students find this topic interesting, and have an easy time of it, so it should not be a hard sell.

POINTS TO ELICIT: Revisit the warm-up to elicit that from a math standpoint, an angle is formed by starting two rays at the same point, and extending them in different directions. The further apart the rays are, the larger the measure of the angle.

DEFINITIONS: The shared point from which an angle's rays originate is called the vertex. The point at which two lines intersect is also called the vertex. Illustrate how four angles are formed by the intersection of two lines.

Explain that angles are measured in degrees. Define a circle as having 360°. If the student asks why, explain that it was just what was established, and the number works out nicely.

Illustrate how a right angle is formed by creating a vertex at the center of a circle, and extending two rays perpendicular to each other. For now just draw them straight up and straight right. The concept of perpendicular is expanded upon later.

Elicit that the resulting angle comprises ¼ of the circle. Since ¼ of 360 is 90, we can infer that our angle measures 90°. Define a right angle as having 90°, and elicit examples of right angles which we see all around us in everyday life.

Illustrate how a straight angle is formed by creating a vertex at the center of a circle, and extending two rays in opposite directions. For now just draw them straight left and straight right. The concept is expanded upon later.

Elicit that the resulting angle comprises ½ of the circle. Since ½ of 360 is 180, we can infer that our angle measures 180°. Define a straight angle as having 180°.

Define an acute angle as measuring greater than 0° and less than 90°. See if the student can deduce that a 0° angle would effectively have its rays on top of each other. Define an obtuse angle as measuring greater than 90° and less than 180°.

POINTS TO REEMPHASIZE: Ensure that the student memorizes the terms from this lesson. Many students assume that definitions will be provided for them during an exam.

IDEAS FOR EMBELLISHMENT: Practice measuring angles using a protractor, and visually estimating the measures of angles. Define a reflex angle as measuring greater than 180° and less than 360° (i.e., a full rotation around a circle).

INSTRUCTOR'S NOTES:

LESSON 30

AIM/TOPIC(S): Complementary and supplementary angles including problems involving algebra

WARM-UP: Draw a complementary angle diagram in which one angle is stated as 37°. Ask the student to deduce the other angle. Then draw a supplementary angle diagram in which one angle is stated as 142°. Ask the student to deduce the other angle. Use any responses to lead into the lesson.

MOTIVATION / DEFINITIONS: Revisit the first warm-up problem. Define the two angles as being complementary which means they add up to 90°. Revisit the second warm-up problem. Define the two angles as being supplementary which means they add up to 180°.

Explain that many test questions involve these definitions, and that some questions will be in the form of a diagram, and some will be in the form of a word problem. In the case of the latter, the definition of complementary and supplementary will not be provided, and therefore must be memorized.

POINTS TO ELICIT: We can use simple subtraction to solve the warm-up problems if we remember that they are based on right (90°) and straight (180°) angles, respectively. It may help to memorize that complementary comes before supplementary alphabetically just as 90 comes before 180 numerically.

DEMONSTRATIVE EXAMPLES: The challenge for this topic is problems which involve algebra. Present this problem either in diagram or word problem format: "Two complementary angles measure $(3x - 11)$ and $(2x + 1)$. Solve for x."

We must set up an equation to solve for x, using what we know. We know that the two angles must add up to 90, so we can set up our equation as $3x - 11 + 2x + 1 = 90$. Combine like terms and algebraically solve for x to get $x = 20$. Obviously if the student has any difficulty at all with the algebra, this lesson must be put on hold while the related material is reviewed.

Explain that the question could have been worded such that instead of asking us to solve for x, we were asked to determine the measure or the smaller or the larger angle. We can do this by first solving for x, and then substituting that value into each expression. In this problem we would get angles of 41° and 49° which we can confirm sum to 90. Obviously we can easily see which angle is the smaller and which is the larger.

A major difficulty which students have with these problems is that they either do not read the question carefully, or they answer with too much information, and end up being marked wrong. For example, if a problem asks for the measure of the larger angle, the student must answer with that and only that. S/he must not include the value of x, and must not include the measure of the smaller angle.

PRACTICE EXERCISE: Present this problem in diagram or word problem format, choosing the opposite format of the previous problem: "Two supplementary angles are represented by the expressions $(3x - 28)$ and $(7x + 58)$. Find the measure of the smaller angle." The problem is solved in the same manner as before, but setting the sum equal to 180. Present related exercises as needed, emphasizing the concepts over rote drilling.

INSTRUCTOR'S NOTES:

LESSON 31

AIM/TOPIC(S): Problems involving parallel lines and a transversal; Vertical angles; More about perpendicularity

WARM-UP: Draw a pair of parallel lines which are not perfectly vertical or horizontal. Then draw another pair of lines which are sloped so that they will eventually intersect. Ask the student to describe the difference between the two pairs of lines. Then draw a pair of perpendicular lines at angles other than perfectly vertical and horizontal. Include the little box in the corner symbolizing perpendicularity. Ask the student what type of angles are formed by the lines.

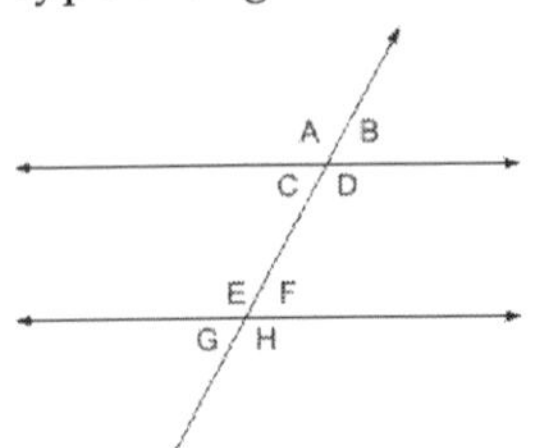

Draw a pair of horizontal parallel lines intersected by a transversal at about 60°. Label the resulting angles A through H, and try to elicit the relationships between them.

MOTIVATION: This lesson involves easy concepts which many students complicate unnecessarily. Try to convince the student by mastering this lesson, many test questions will be easy.

POINTS TO ELICIT: Revisit the parallel lines from the warm-up. Elicit that the lines will never intersect. We say that such lines are parallel. Ensure the student understands that parallel lines need not be horizontal or vertical.

Revisit the perpendicular lines from the warm-up. Elicit that four right angles are formed, and that the little square in the corner indicates that the lines are perpendicular. The lines need not run perfectly vertical and horizontal, as long as they intersect to form right angles. Review that if two angles sum to a 90° right angle, we say that they are complementary.

Revisit the warm-up problem with the parallel lines and transversal. There only two concepts to elicit. The first is that in such a setup, all of the small angles are equal in measure to each other, and all of the large angles are equal to each other. The second concept is that the sum of any small angle plus any large angle is 180°. This should be reminiscent of the supplementary angle problems from the last lesson. Elicit that as long as we are given the measure of any angle in the diagram, we can easily deduce the measure of any other.

ANGLE RELATIONSHIP NAMES: In the diagram at left, the most common angle relationship is that of angles A-D or B-C. Those pairs are called vertical (opposite) angles. Elicit that two vertical angles are equal in measure, but not necessarily equal in measure to two other vertical angles. If required for an exam or coursework, teach the names given to other pairs of angles in the diagram. A-E are corresponding, C-F are alternate interior, and B-G are alternate exterior. Have the student find the other pairs in each relationship.

DEMONSTRATIVE EXAMPLES: Some transversal problems involve algebraic expressions just like the last lesson. If a problem involves a small and a large angle, set up an equation in which the two expressions are summed and set equal to 180. If a problem involves two small angles or two large angles, set the two expressions equal to one another. Emphasize that the student must answer the specific question asked.

INSTRUCTOR'S NOTES:

CHAPTER EIGHT

Lesson Plans for "Area, Perimeter, and Volume"

Lessons 32 to 40

Topics Covered in This Chapter:

Perimeter; Introduction to area; Square units; Area of a rectangle, square, triangle, parallelogram, and trapezoid; Circle terms; Introduction to pi (π); Area and circumference of a circle; Introduction to volume; Cubic units; Volume of a rectangular solid, right circular cylinder, triangular prism, cone, square pyramid, and sphere; The effect on area/volume when increasing dimensions

LESSON 32

AIM/TOPIC(S): Perimeter; Introduction to area; Square units; Area of a square and rectangle

WARM-UP: Ask the student where s/he has heard the words area and perimeter in everyday life (e.g., area rug, military perimeter, etc.), and assess his/her intuitive understanding of the concepts. Use any response to lead into the lesson.

MOTIVATION: This lesson introduces the basic building blocks of geometry. If it is skipped or rushed, the student will have difficulty with all the geometry that is to follow.

POINTS TO ELICIT: Most students have an intuitive understanding of the perimeter concept. Ensure that the student understands to only add up the sides which comprise the distance around the given figure (polygon), and that s/he will not also add in any decoy data such as internal measurements.

We must sometimes deduce the lengths of missing sides, but at this level it will be easy. Draw a parallelogram with the measures of the left and top sides noted. Elicit we can deduce the measures of the other two sides to compute the perimeter.

Most students have an intuitive understanding of area as the space inside a shape. However, most students also harbor many misconceptions on the topic. The main concept to instill is that area is based upon two perpendicular dimensions which we could think of as length and width. This is easiest to visualize for a rectangle, but of course it applies to any two-dimensional shape. A more complicated shape like a trapezoid has a more elaborate area formula, but it is still based upon only two dimensions. Many students think that the area of a triangle involves three dimensions since it has three sides.

Area is always based on how many unit squares can fit inside a given shape. Again, this is easiest to visualize for a rectangle, but it applies to any two-dimensional shape. Many students aren't comfortable with the idea of filling a right triangle with unit squares, only to have to "cut" the ones that lie along the slanted side. Students are even more uncomfortable with the idea of filling a circle with unit squares, knowing that the ones along the edges would have to be cut and rounded. It is difficult for most students to believe that if you were to take all of the cut and/or rounded partial squares which comprise a shape's area, and assemble those parts together like a puzzle, the area would work out correctly.

DEMONSTRATIVE EXAMPLES: Draw a 7 in. by 3 in. rectangle (actual size). Show how it can be filled with 21 squares which are each 1 in. by 1 in. (i.e., unit squares). Explain that we must represent the area as 21 in^2 (read as "21 square inches," and sometimes written as 21 sq. in). It is wrong to write the answer as 21 in. which would be a length and not an area. Repeat the above exercise for a 6 cm. by 4 cm. rectangle to show that the area is now based on square centimeters. Elicit that the area formula for a rectangle is length times width.

Repeat the above exercises for squares of various sizes and units, and elicit that while the formula is still effectively length times width, in the case of a square we can write the area formula as $A = s^2$ since length and width are equal.

INSTRUCTOR'S NOTES:

LESSON 33

AIM/TOPIC(S): Area of a triangle

WARM-UP: Draw a right triangle of height 3 in. and base 5 in. Don't label the hypotenuse. Elicit the student's ideas on how to compute the area. Use any response to lead into the lesson.

MOTIVATION: While it is practical to measure the area of rectangles, we don't encounter many triangles in everyday life. Still, problems involving the area of triangles are common and easy if a few key concepts are understood.

POINTS TO ELICIT / DEMONSTRATIVE EXAMPLES:
Revisit the warm-up problem. Remind the student that we learned two main concepts—area is always based on two perpendicular dimensions, and is always based on square units. We fill a shape with unit squares, even if some of them would need to be cut along slanted or curved edges.

Ask the student how the warm-up triangle is related to a 3 in. by 5 in. rectangle. Try to elicit that we could extend lines to show that the triangle is half of such a rectangle. Knowing that the area of that rectangle is 15 in^2, elicit that the area of the triangle must be 7.5 in^2, but the reason may not yet be known.

Ask the student to fill in the triangle with unit squares, starting at the right angle. S/he should be able to fit in three without issue. A fourth one would fit in along the bottom if a bit was cut off along the slanted edge. Have the student experiment in this fashion until s/he is convinced that if the "cut off" parts of squares were used to fill in the parts of the triangle which weren't filled, an area of 7.5 square inches seems right.

The student may ask how s/he can be certain that the area is 7.5 sq. in. since it's hard to accurately assemble the partial squares, especially if not actually cutting them out. Acknowledge that this is why we use formulas in geometry. Once the formula is established, we just have to understand and memorize it, and it need not be "reinvented."

Many students will insist that we must fill a triangle with "unit triangles," or at least write our answer in those terms. Continue to stress that area is always represented in square units.

Elicit that since the area of a rectangle is length times width, and our triangle's area worked out to be half that, we can begin to deduce the formula for the area of a triangle. However, some important modifications must be made.

Draw a triangle whose height is dropped to a (dotted line) extension of the base. Draw another triangle whose height is dropped within the triangle itself. Explain that when computing the area of a triangle, instead of length and width, we use measures of base and height, with height always measured from the highest point of the triangle straight down (perpendicular) to the base (or an extension of such). Continue to stress this important point, and ensure that the student knows to ignore any decoy data such as the measures of slanted sides.

Elicit that the formula for the area of any triangle is $\frac{1}{2}bh$. Many students cannot accept that this works for any triangle, so do your best to drive the point home. Ensure the student understands that if the base needs to be extended to drop the perpendicular height, the extension is not part of the base.

INSTRUCTOR'S NOTES:

LESSON 34

AIM/TOPIC(S): Area of a parallelogram and trapezoid

WARM-UP: Draw a parallelogram of base 5 in. and height 2 in. Draw the slanted sides so that they are about 3 in. Label them as such, and explain the concept of "not drawn to scale."

Then draw a trapezoid with bases of 5 in. and 7 in., and height of 2 in. Draw the slanted sides so that one is about 3 in. and the other is about 4 in., and label them as such. Ask the student for his/her ideas on what the area of each figure is. Use any responses to lead into the lesson.

MOTIVATION: Problems with parallelograms and trapezoids are common and easy if a few key concepts are understood.

POINTS TO ELICIT / DEMONSTRATIVE EXAMPLES:
Revisit the warm-up parallelogram. Elicit that had the figure been a 5 in. by 2 in. rectangle, the area would have been 10 in^2. Informally, our parallelogram is just a "slanted" rectangle, so it makes sense that the area would be similar, if not the same.

Drive home the point that the slanted sides do not factor into the formula at all. Elicit that the formula for the area of a parallelogram is bh, where h (height) is the distance between the two bases, measured perpendicular to each. Discuss the properties of a parallelogram, keeping in mind that at this level the student need only understand the general concept.

Revisit the trapezoid from the warm-up. Elicit that it's not unreasonable to average the two bases, and assume that each one was 6 in. Had the figure been a 6 in by 2 in. rectangle, the area would have been 12 sq. in. Informally, our trapezoid (with the bases averaged out) is just a variant of a rectangle, so it makes sense that the area would be similar in nature.

Drive home the point that the slanted sides do not factor into the formula at all. Elicit that the formula for the area of a parallelogram is $\frac{1}{2}(b_1 + b_2)h$, where h (height) is the distance between the two bases, measured perpendicular to each. Elicit that this formula generates the average of the two bases since we're effectively adding them and dividing by 2. Sometimes the formula is represented as such. Discuss the properties of a trapezoid, keeping in mind that at this level the student need only understand the general concept.

PRACTICE EXERCISES: Make up additional similar practice exercises involving parallelograms and trapezoids, only to the extent necessary to drive home the key concepts and formulas. There is no point in endlessly drilling.

POINTS TO REEMPHASIZE: Ensure the student understands that in these shapes (as well as triangles), the height is always measured perpendicular to the base. Most students do not feel comfortable disregarding decoy data which commonly appears in diagrams. Ensure the student understands how the trapezoid area formula works. Many students will multiply the bases, or simply say that they forgot the formula. Elicit that the formula need not even be memorized as long as it is understood. The formula tells us to average the two bases (which the student should already feel comfortable doing), and multiply that result times the perpendicular height.

INSTRUCTOR'S NOTES:

LESSON 35

AIM/TOPIC(S): Circle terminology; Pi (π)

WARM-UP: Ask the student where s/he encounters circles in everyday life. Survey the student to see what ideas s/he has about pi (π), if any. Use any responses to lead into the lesson.

MOTIVATION: Circles and pi (π) play a big role in geometry. It is essential for the concepts of this lesson to be mastered. Most students don't have a hard time with this lesson, but often struggle with getting past their misconceptions about pi.

POINTS TO ELICIT / DEMONSTRATIVE EXAMPLES:
Ask the student how we could go about measuring the distance around a circle. Try to elicit that we could get an approximate value by laying string along the circumference, and then using a ruler to measure how much string we used. Do this if time permits. Explain that the distance around a circle is called the circumference instead of the perimeter.

If time and materials permit, have the student use a compass to draw a circle with a radius of 2 inches. Define the radius as the distance from the center of the circle to any point along the circle's edge. Dispel the misconception that a radius can only be horizontal. Define the diameter as twice the radius, and elicit that a diameter runs from one edge of the circle to the other, passing through the exact center. Elicit that a diameter need not be horizontal, and that a circle has an infinite number of diameters and radii (plural). Define a chord as a line which starts at a point on the circle and ends at some other point on the circle, but does not pass through the center.

Have the student use string to measure the circumference of our 4 in. diameter circle. It should be about 12.5″. If measuring is not practical, just state it and ask the student to "trust you." Have the student take the circumference and divide it by the diameter. S/he should get a value which is a bit larger than 3.

Have the student repeat the above exercise but using a circle with a 5″ diameter. The result of computing C/d should again be a bit larger than 3. If time permits, repeat the exercise using a circle with a 3″ diameter. Again, the result of C/d should be the same. Try to convince the student that regardless of a circle's size is, the result of computing C/d is bit larger than 3.

Explain that if we had more precise measuring devices, the result of C/d would always be 3.14 if we rounded it to the hundredths place, but that the decimal places of the actual value continue infinitely without repetition or pattern. We call this value the Greek letter pi (π). Define π as the ratio of the circumference to the diameter in any circle. Elicit that π itself does not have any units. When we divide the units of C by the units of d, the units (such as inches) "cancel."

Pi is an irrational number. Informally, this means that it cannot be represented as a fraction or as a terminating decimal. With that said, explain that it is common to use 3.14 as an approximate value of pi in computations. We also sometimes use 22/7. However, stress that both of those values are precisely that—approximations. Many test questions are designed to see if the student understands that concept. The instructions to a problem will indicate what value to use, and how to round the final answer. Reiterate that π is extremely important in math, and should not be dismissed as trivia.

INSTRUCTOR'S NOTES:

LESSON 36

AIM/TOPIC(S): Area and circumference of a circle

WARM-UP: Elicit the student's ideas on how we might compute the area and circumference of a circle. Answers will vary, but will help you to determine whether previous lessons on area must be reviewed, and/or if the student is harboring any misconceptions. Use any responses to lead into the lesson.

MOTIVATION: Computing the area and circumference of a circle are very common tasks in geometry. They are simple as long as a few key concepts are understood.

POINTS TO ELICIT: In the last lesson we learned $C/d = \pi$. We know the value of π, so if we want to compute C, we will have to either be given the value of d, or we will need enough information to deduce it (e.g., the value of the radius). To compute C, we will have to isolate it on one side of the equation. Elicit that we can do this by multiplying both sides by d. Review the lessons on basic algebra if necessary. We are left with $C = \pi d$ which is the formula for circumference (a length). Elicit that $C = 2\pi r$ is an equivalent formula.

We learned that area is always based on two perpendicular dimensions, but in a circle, we cannot really speak of length and width as we could with a rectangle. But, we can do something involving the diameter (or the radius).

Present the formula for the area of a circle: $A = \pi r^2$, and instruct the student to memorize it. The formula can often be a tough sell, especially when students insist on a proof which is far too complex at this stage. The following usually serves as an effective response: Draw a circle with a horizontal radius and a vertical radius. Elicit that the formula serves to multiply those two radii together. In a sense, we could think of this as multiplying one dimension times the other, just like we did with a rectangle. Some students will ask why pi is part of it, or why the formula doesn't involve d^2. At this level of math it is hard to provide a convincing answer.

The biggest problem students have is confusing the formulas for circumference and area. Elicit that the circumference formula is just an algebraic manipulation of what we learned in the previous lesson, and circumference is always given in terms of length (i.e., units—not units squared). A formula involving squaring is certainly wrong. The area formula makes sense because it results in an answer with squared units.

Elicit that when using the area formula we end up with square units as required. Remind the student that pi itself has no units, and that if we square the radius (which is a length), we end up with a length squared (e.g., in^2 or sq. in.).

PRACTICE EXERCISES: At this level, problems will involve computing C when d (or r) is given, and computing A when r (or d) is given. The student is typically instructed to use 3.14 as an approximate value of pi, and to round his/her final answer to the nearest tenth or hundredth. Make up a variety of practice problems as needed, prioritizing concept comprehension over rote, repetitive exercises. Ensure that the student includes the units of length when computing circumference, and squared units when computing area.

INSTRUCTOR'S NOTES:

LESSON 37

AIM/TOPIC(S): Introduction to volume; Cubic units; Volume of a rectangular solid

HELPFUL MANIPULATIVES: For the next few lessons on volume, it is helpful to have some tangible 3-D figures for the student to examine. These are commercially available. If nothing else, an empty toilet paper roll is an example of a cylinder, and a box is an example of a rectangular prism.

PREREQUISITE KNOWLEDGE: It is essential that the student be fully comfortable with the lesson on area and square units, as well as the lesson which introduced geometric dimensions.

WARM-UP: Have the student draw a rectangle on the board (or on paper), and mentally "extend it outward" from the surface. Elicit what type of figure results. The student may have a hard time with visualization, but don't allow him/her to get discouraged, or feel as though s/he will not be able to handle this unit on volume. Have him/her repeat the exercise with a square, but this time extending it the same amount of distance as one of its sides. Elicit what figure results. Ask the student to think of real-world examples of each 3-D figure.

MOTIVATION: This lesson is just an extension of the lesson on area and square units. Many students overly complicate the matter, in some cases fearing that they will be required to draw representations of 3-D solids. Sometimes students have a hard time with the fact that we live in a 3-D world, but on paper we are forced to make 2-D representations of 3-D objects. Try not to let the student get overwhelmed.

POINTS TO ELICIT / DEMONSTRATIVE EXAMPLES:

Revisit the rectangle warm-up task. We get what is commonly called a box. In geometry it is called a rectangular solid or rectangular prism. We refer to 3-D objects as solid, even though they are not necessary "filled." Our usual task is to compute the space inside, known as the volume. Elicit that this is analogous to computing the area for 2-D shapes. Revisit the second warm-up task, and elicit that we get a cube.

Since we measured area in terms of how many square units could fit inside the 2-D shape, it follows that to measure volume, we would determine how many unit cubes could fit inside the 3-D figure. This is difficult to represent via a 2-D drawing, but it is really not necessary to try. The student just needs to understand that we fill 2-D shapes with unit squares, and 3-D shapes with unit cubes. The answers to volume problems are always given in terms of cubic units ($units^3$).

Have the student think about an empty box that is 12 by 5 by 8 inches. To compute the volume, we will fill the box with unit cubes measuring 1 in. × 1 in. × 1 in. Elicit that it will take 480 cubes to fill the box, derived from multiplying the given dimensions. The volume is 480 in^3. Have the student try to visualize this, but again, emphasize that this is just the 3-D extension of multiplying length times width to compute the area of a rectangle. Elicit that the formula for the volume of a rectangular solid is length × width × depth (or height).

The volume of a cube uses the same formula, but in a cube, all the dimensions are equal. We instead use the formula $V = e^3$, where e is the length of an edge of the cube. Elicit that we will be filling a large cube with smaller unit sized cubes.

INSTRUCTOR'S NOTES:

LESSON 38

AIM/TOPIC(S): Volume of a right circular cylinder and triangular prism

PREREQUISITE KNOWLEDGE: Review the previous lesson and the lesson on area of a triangle and circle.

WARM-UP: Have the student draw a circle on paper or on the board, and visualize extending it outward just like we did with the figures in the last lesson. Elicit the student's ideas on what figure results, and where s/he sees that figure in everyday life.

Have the student repeat this task but using a triangle as the base. Any type of triangle will be fine. For this task, just try to have the student visualize what will result if the triangle is "extended straight outward" from the surface.

MOTIVATION: This lesson extends the unit on volume to two new solids which are common in geometry.

POINTS TO ELICIT: Revisit the first warm-up problem. Elicit that we get a cylinder. It is more accurately described as a right circular cylinder since it has a circular base, and is extended perpendicular to its surface, as opposed to "leaning."

Try to elicit the volume formula for a cylinder. Although we will still fill it with unit cubes, it will be very hard to visualize doing so, just as it was hard to visualize filling a 2-D circle with unit squares. Instead, we will just extend the formula for the area of a circle into the third dimension. The main book has a diagram which may be helpful to examine at this time, as would be any manipulatives described in the last lesson.

We know that the formula for the area of a circle is πr^2. To extend it to the third dimension, all we do is multiply it by the third dimension which is usually called height or depth. The formula for the volume of a cylinder is $\pi r^2 h$. This is analogous to the formula for a rectangular solid which is just the area of a rectangle multiplied by the third dimension.

Revisit the second warm-up problem. This one is a bit trickier. We get what is commonly called a triangular prism. The bases on the "end" faces are triangles. See if you can elicit that each other face is actually a rectangle. Have the student try to deduce the volume formula as above. The area of the base triangle is ½ *bh*, with both of those variables having nothing to with the third dimension. We must then multiply that resulting area times the third dimension.

While is it least confusing to refer to that third dimension is depth, unfortunately many textbooks and teachers refer to it as height, and abbreviate it h. It is essential to stress that the third dimension of a triangular prism is unrelated to the height of the base triangle. Present the volume formula as $\frac{1}{2}\,bh \times d$, but again emphasize that if the student sees d represented as h, it is a totally different h than the height of the triangle.

PRACTICE EXERCISES: The main book has some practice exercises to work with, but as always, ensure that the concepts are being grasped. Stress that the theme of the volume lessons is really just "two dimensions times the third dimension."

POINTS TO REEMPHASIZE: Remember, volume is in given in cubic units, not prism units or cylindrical units or similar.

INSTRUCTOR'S NOTES:

LESSON 39

AIM/TOPIC(S): Volume of a cone, square pyramid, and sphere

WARM-UP: Ask the student where s/he has encountered cones, pyramids, and spheres in everyday life. Use any response to lead into the lesson.

MOTIVATION: This short lesson teaches how to compute the area of some other solids which are less common than the ones we already worked with. It serves as a good review. Note that some exams may provide the formulas which are taught in this lesson, but if not they should be memorized.

POINTS TO ELICIT: A cone could be thought of as a tapered cylinder, so its volume formula is related. The volume of cone is given by $\frac{1}{3}\pi r^2 h$ which is 1/3 the volume of a cylinder.

A sphere is the 3-D counterpart to a circle so its volume formula is related. It is given by $\frac{4}{3}\pi r^3$. The cubing of the radius makes sense since we're dealing with a 3-D figure. The 4/3 part just needs to be memorized.

A square pyramid has a square base, and four triangular sides which meet at a point. Its volume formula is given by $\frac{1}{3}s^2 h$ where s is the side of the square base. In some textbooks, s^2 is represented as B, where B is the area of the square base.

PRACTICE EXERCISES: The main book has some practice exercises and 2-D representations of these solids, but time may be better spent reviewing previous material.

INSTRUCTOR'S NOTES:

LESSON 40

AIM/TOPIC(S): The effect on area/volume when increasing dimensions

PREREQUISITE KNOWLEDGE: Review all of the previous material on area and volume

WARM-UP: Ask the student to imagine a rectangle of any convenient dimensions. Then ask how the area would be affected if we were to double the length, and double the width. Use any response to lead into the lesson.

MOTIVATION: This quick lesson teaches how to solve a very popular question model on geometry exams. It is very easy as long as one simple concept is understood. It is best to use any extra time to review earlier material or move ahead.

POINTS TO ELICIT / DEMONSTRATIVE EXAMPLES:
Revisit the warm-up problem. Many students say the area will double since the dimensions have doubled. Try solving the problem using any starting measurements, for example 3 by 5. It is easy to see that doubling each dimension results in an area which is quadrupled. Elicit that this makes sense since the area formula is length times width. We multiplied a doubling times a doubling.

Ask the student how the area would be affected if we multiplied one dimension times two, and the other times three. Elicit that the area would be 6 times as big (doubling times tripling).

Elicit that this concept is the same for a triangle, since the area of a triangle is computed by multiplying two dimensions. The fact that the formula involves ½ of that result doesn't change matters. If we triple the base and triple the height, the resulting area is multiplied by 9.

The concept is the same for a circle as well, but it's slightly different. What happens if we double the radius? Elicit that the formula for the area of a circle involves squaring the radius, which means multiplying radius times radius. Whatever alteration we make to the radius ends up getting squared. If we double the radius, the area quadruples. If we triple the radius, the area becomes 9 times as big, and so on.

The concept can be extended to the volume of 3-D solids. Volume formulas all involve multiplying three dimensions. If one or more of those dimensions are increased, we just multiply all of the increases together to determine the effect on the volume. For example, if a rectangular solid has its length doubled, its width tripled, and its height quadrupled, the resulting volume will be $2 \times 3 \times 4$ or 24 times as big.

As another example, if we multiply the radius of a sphere times 4, the resulting volume will be 4^3 or 64 times as big.

For more practice, elicit how the volume of a cone is affected if we quadruple the height and triple the radius. Remember, the formula calls for the radius to be squared (i.e., multiplied times itself). The resulting area is 4×3^2 or 36 times as big.

PRACTICE EXERCISES: The main book has some practice exercises. Ensure the student grasps the one main concept presented in this lesson—all changes to a figure/solid get multiplied together to determine the increase on area/volume.

INSTRUCTOR'S NOTES:

CHAPTER NINE

Lesson Plans for "The Pythagorean Theorem and Triangles"

Lessons 41 to 44

Topics Covered in This Chapter:

Sum of internal angles in a triangle; Types of triangles; Algebra problems involving triangles; Similar and congruent triangles; Word problems which lead to a problem of similar right triangles; The Pythagorean Theorem; Pythagorean triples

LESSON 41

AIM/TOPIC(S): Sum of internal angles in a triangle; Types of triangles

CONNECTIONS TO EARLIER/LATER MATERIAL:
This quick lesson introduces some important concepts and definitions which will be used throughout the rest of the unit. Use any extra time to review previous material or move ahead once this lesson is mastered.

WARM-UP: Ask the student to state how many degrees are in each angle of a rectangle, and then to sum those measurements. Remind him/her that in an earlier lesson, we formed a triangle by "cutting" a rectangle in half diagonally. Elicit the sum of the angle measurements in a triangle. Ask the student for his/her her ideas on whether every triangle's angles sum to 180 degrees. Use any response to lead into the lesson.

MOTIVATION: This lesson introduces the unit on triangles and the Pythagorean Theorem—two very popular topics on exams.

POINTS TO ELICIT / DEMONSTRATIVE EXAMPLES:
Explain that the sum of the (internal) angles in any triangle is 180°. If there is time, it is ideal if the student can experiment by drawing triangles and measuring the angles with protractor, or using a computer-based program for this. Most students can accept that as one angle gets bigger, at least one of the other two must get smaller, maintaining a consistent sum of 180°.

Many problems involve nothing more than this fact. If two angle measurements in a triangle are stated, the third angle measurement can be computed just by subtracting the first two from 180. Don't allow the student to complicate the matter.

It is important for the student to be familiar with several common types of triangles. The first triangle below is an equilateral triangle. It has three equal sides and three equal angles. Demonstrate the proper use of side and angle tick marks. Elicit that each angle must be 60°.

The second triangle below is an isosceles triangle. It has two equal sides, and one unequal side. The base angles that are opposite the two equal sides are equal to each other. The vertex angle at the top is unequal. We can use our knowledge of these properties to solve problems which we will see later.

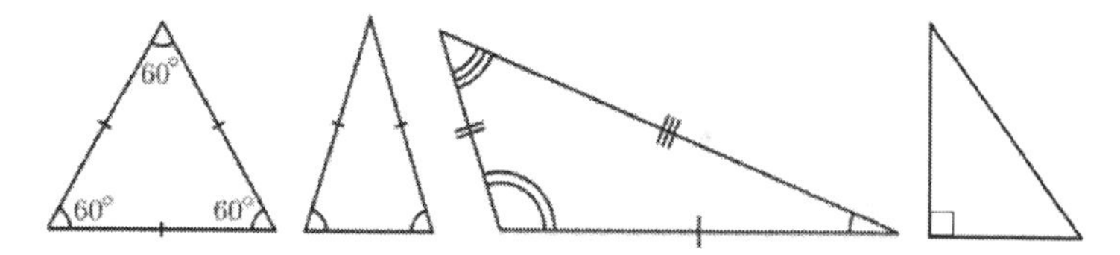

The third triangle above is a scalene triangle. It has three unequal sides and three unequal angles as denoted by the side and angle tick marks.

The fourth triangle above is a right triangle which we've already worked with. It contains a 90° right angle which means that the other two angles must sum to 90°. Elicit the characteristics of a right isosceles triangle. It would have a 90° right angle. To satisfy the conditions of an isosceles triangle, the other two angles would each have to be 45°. Ensure that the student memorizes the definitions from this lesson.

INSTRUCTOR'S NOTES:

LESSON 42

AIM/TOPIC(S): Algebra problems involving triangles

PREREQUISITE KNOWLEDGE: This lesson relies on concepts taught in the lessons about angles and all the lessons involving algebraic techniques. It serves as a good review of those topics. If the student has difficulty with this lesson, you will need to assess whether it is due to an issue with the geometry, or if the required algebra was never fully mastered.

WARM-UP: Present this problem to think about: A triangle's angles are given by the expressions or constants 5x+7, 2x–9, and 42. Find the measure of the smallest angle. Elicit the student's ideas on how to proceed, and optionally allow him/her to do so. Use any response to lead into the lesson.

MOTIVATION: This quick lesson teaches how to solve triangle problems using algebra. Such problems are very common on standardized exams since they incorporate many concepts.

POINTS TO ELICIT / DEMONSTRATIVE EXAMPLES:
Revisit the warm-up problem. Elicit the solving process. We know the sum of the angles will be 180. We can create an equation in which we sum the given expressions or constants, and set it equal to 180. We can then solve for x using the techniques we learned, and substitute that value in each expression involving x to determine the measure of each angle.

Ask the student to recheck what is being asked for. We must state the measure of the smallest angle. It is wrong to answer with the largest angle, or to answer with the value of x. We must answer the specific question, and only that question.

Elicit that our equation is $5x + 7 + 2x - 9 + 42 = 180$. Solving algebraically for x we get $x = 20$. Substituting in the original expressions we see that the angles measure 107, 31, and 42. Confirm they add up to 180. The answer is 31°. Any other answer is wrong, as is including any additional information.

For more practice, present this problem: An isosceles triangle has base angles represented by the expressions $8x + 16$ and $2x + 28$. Solve for x. In this case we must remember that the two base angles are equal. We can equate their expressions in the equation $8x + 16 = 2x + 28$. Solving algebraically we get $x = 2$. Any other answer is wrong. Check via substitution to see that both expressions have the same value.

PRACTICE EXERCISES: The main book has some additional exercises. It is important for the student to understand that a wide variety of questions involving algebra and triangles can be asked. To be prepared for them, it is essential to have fully mastered the algebraic techniques presented in this book. It is also essential to know the attributes of the different types of triangles which were presented in this chapter.

POINTS TO REEMPHASIZE: Many students do not read questions carefully. Do not assume that a problem is asking you to solve for x. Many students also think that they will get extra credit by providing additional information, such as the measurements of all angles even if the question only asked for the measure of the smallest or largest angle. Explain that not only will the student not get extra credit for providing additional information, his/her answers will be marked wrong.

INSTRUCTOR'S NOTES:

LESSON 43

AIM/TOPIC(S): Similar and congruent triangles, including related word problems

PREREQUISITE KNOWLEDGE: Review the lesson on how to solve proportions algebraically, and ensure that it is mastered.

WARM-UP: Present the diagrams below. Ask the student why s/he thinks that these triangles are considered to be similar. Elicit his/her ideas on how we could use proportions go about solving for x and y. Use any responses to lead into the lesson.

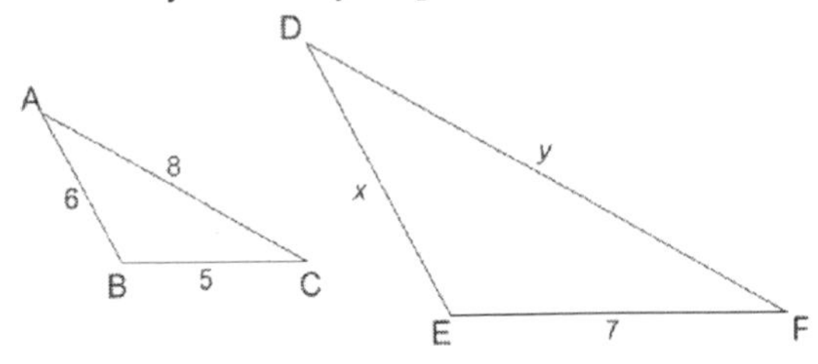

MOTIVATION: This lesson teaches how to solve diagram and word problems involving similar triangles. It is a topic which appears frequently on exams because it relies on several different concepts.

POINTS TO ELICIT / DEMONSTRATIVE EXAMPLES:
Revisit the warm-up problem. By stipulating that the two triangles are similar, we are saying that their respective sides are in proportion. Symbolically we can write $\Delta ABC \sim \Delta DEF$. Informally, we can say that one triangle is an enlargement of the other. In math, the word similar has this precise meaning.

To solve for x, we must set up a proportion using corresponding sides. One method is $5/7 = 6/x$. Informally we can say "base is to base, just as left side is to left side." We could have also set up the proportion as 5/6 = 7/x (base is to left just as base is to left). Cross-multiply to get x = 8.4. Always check the answer for reasonableness. For practice, have the student set up a proportion to solve for y. The answer is y = 11.2.

During this lesson the student should learn what the term congruent means. In math, if two shapes are congruent, they are exactly the same size and shape. Informally we could think of the figures as exact photocopies of each other.

To review, if two figures are similar, one could be thought of as as enlargement of the other, but all of the sides are still in proportion, and all angle measurements are maintained. It is important to understand that rotation does not affect congruence or similarity. A trick question may rotate a figure, but the rotation should not be taken into account.

Many problems involving similar triangles start out as word problems. At this level they always follow the exact same format. They involve two adjacent things such as flagpoles, people, or trees, each casting a shadow.

The only challenge in these problems is translating the words into a diagram which will be two similar right triangles. The problem is then solved just as above, namely setting up a proportion involving the missing side.

PRACTICE EXERCISES: The main book has additional practice exercises, but problems will always follow this same format. Ensure that the main concepts have been mastered.

INSTRUCTOR'S NOTES:

LESSON 44

AIM/TOPIC(S): Pythagorean Theorem.; Pythagorean Triples

PREREQUISITE KNOWLEDGE: This important lesson relies heavily upon knowledge of squaring, perfect squares, square roots, right triangles, and basic algebraic equations. The lesson will be a struggle for both you and the student if those topics have not been fully mastered.

WARM-UP: Draw a right triangle with legs measuring 3 inches and 4 inches. Define those sides as legs. Tell the student that

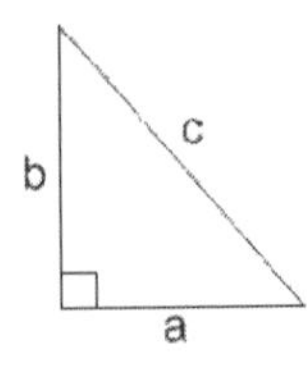

the length of the unlabeled longest side, called the hypotenuse, can be computing using the formula $a^2 + b^2 = c^2$ where a is the length of one leg, b is the length of the other, and c is the length of the hypotenuse. Ask the student to use the formula and his/her knowledge of algebra to compute c. Use any response to lead into the lesson.

MOTIVATION: This three-page lesson introduces the Pythagorean Theorem. It is extremely popular among test-makers, so it is important for the student to become proficient with it.

POINTS TO ELICIT: Revisit the warm-up. Explain that the formula you provided is called the Pythagorean Theorem. Stress that it applies only to right triangles. If we know the lengths of two of the three sides of a right triangle, we can use the Theorem and basic algebra to compute the unknown side.

After substituting and combining like terms we have $25 = c^2$. Many students will write 25 as the answer. Elicit that 25 couldn't possibly be correct based on a quick visual estimate, as well as what we know about triangles in general. If two sides are 3 and 4 units, the third side would have to be smaller than 7 units otherwise the shape wouldn't be closed. Give the student time to think and/or draw diagrams until convinced.

To compute c, we must take the square root of each side. Stated another way, we must determine what number squared gives us 25. One answer is 5, but elicit that -5 is also a correct answer since $(-5)^2 = 25$. Explain that in geometry problems involving measurements, we just discard any negative solutions since lengths are always positive.

DEMONSTRATIVE EXAMPLES: Present this example either in diagram or word problem format. Ensure that the student becomes comfortable with both: A right triangle has a leg of 5 ft. and hypotenuse of 13 ft. What is the length of the other leg?

Elicit that we can use the Theorem, but we must understand that the unknown side is not c. We can set up our equation as $5^2 + b^2 = 13^2$. After squaring we have $25 + b^2 = 169$. Subtract 25 from each side to get $b^2 = 144$ which yields b = 12 ft.

Sometimes we will use the Pythagorean Theorem, and end up with something like $b^2 = 48$ which is not a perfect square. At this level what we would do is take the square root of each side to get $b = \sqrt{48}$. We would likely be instructed to compute the square root using a calculator, and rounding our answer to a given decimal place value. In later math we will learn other ways of representing that answer.

Explain that when we use the Pythagorean Theorem, we often end up with a square root answer as above. However, some right triangles have sets of sides (leg-leg-hypotenuse) whose numbers work out "nicely." These sets of numbers are called Pythagorean Triples. By memorizing them, we won't have to "reinvent the wheel" and use the Theorem when we encounter a problem which involves them.

The most common Triple is 3-4-5. Explain that any multiple of those numbers is also a Triple, such as 6-8-10. The second-most common triple is 5-12-13 and its multiples. Other Triples are much less common, and may not be worth memorizing.

Here is a problem which involves a Pythagorean Triple. A right triangle has a leg of length 15 units, and a hypotenuse of 39 units. What is the length of the unknown leg? Examination will show that this does not follow the pattern of a 3-4-5 triangle or its multiples. It is, however, a multiple of a 5-12-13 triangle. It is three times as big. This means that the missing leg must be three times as big as 12 units, which is 36 units. There was no need to even use the Theorem.

For practice, have the student substitute the two common Pythagorean Triples and some of their multiples into the Theorem until convinced that they are valid.

Ensure that the student will not get tricked into the use of a Triple. For example, if a right triangle has legs of 3 and 5 units, that does not mean that the hypotenuse will have 4 units. Let the student sit with that until convinced. For such a problem we must use the Theorem, and we'll get a square root answer.

PRACTICE EXERCISES: The main book has additional practice exercises, but as always, just ensure that the formula is memorized, and the concepts are fully grasped.

INSTRUCTOR'S NOTES:

CHAPTER TEN

Lesson Plans for "Linear Equations and the Coordinate Plane"

Lessons 45 to 50

Topics Covered in This Chapter:

Introduction to the Cartesian coordinate plane; Plotting points; Computing slope using the slope formula; Zero and undefined slopes; Y-intercept; Linear equations in slope-intercept form; Converting from standard to slope-intercept form

LESSON 45

AIM/TOPIC(S): Introducing the Cartesian coordinate plane

REQUIRED/SUGGESTED MATERIALS: For all the lessons in this chapter, it is essential for the student to have graph paper. It is common for students to draw makeshift coordinate planes on lined or blank paper which is highly discouraged. Of course you will need graph paper as well. If you use a dry erase board or easel, consider using a flip-chart of 1″ graph paper. Multicolored pencils or markers can be put to good use if available.

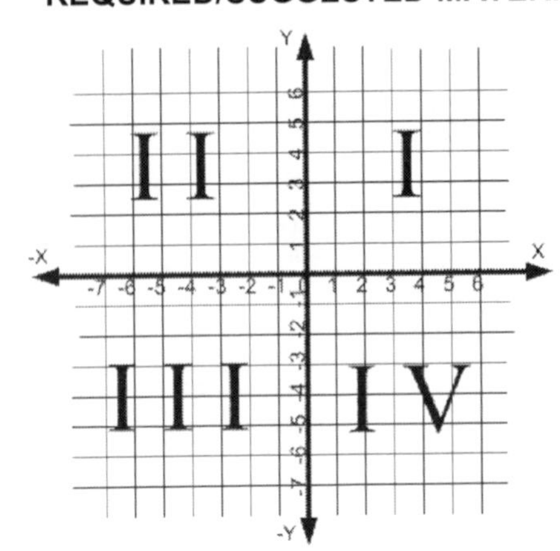

PREREQUISITE KNOWLEDGE: For all of the lessons in this chapter, the student will need to feel very comfortable with signed number arithmetic, especially that which involves subtraction of negatives. Review previous material as needed.

WARM-UP: Draw a Cartesian coordinate plane as shown, and have the student copy it using his/her graph paper. There is not much in this quick lesson for the student to actually "do." Plan on introducing the concepts and terminology of the lesson, and then quickly moving ahead to the next lesson on identifying and plotting points on the coordinate plane.

MOTIVATION: This lesson begins the unit on the Cartesian coordinate plane which is a large component of geometry. The topic is not hard at this level, but many students complicate matters unnecessarily. Explain that this unit teaches some introductory concepts and terminology. Much of it will make more sense in later lessons in the unit.

POINTS TO ELICIT: We will examine the attributes and characteristics of the coordinate plane one by one. Have the student observe the x and y-axes. Elicit that the axes are really just number lines with positive and negative sides. The x-axis runs horizontal, and the y-axis runs vertical. Explain that this is always how we always set up a graph. Have the student observe that the two axes intersect at their respective 0 points. We call this the origin. It will be elaborated on later.

The axes divide the coordinate plane into four quadrants numbered as shown. Have the student memorize the quadrant numbers which are represented with Roman numerals.

Explain the concept of scaling the graph, which at this level is done by ones. Ask the student where s/he has seen graphs which were scaled differently (e.g., charts in news articles, etc.) Explain that a graph must always be scaled, and have its axes labeled. Points are usually lost unnecessarily if they aren't. The quadrants are typically not labeled, though.

PRACTICE EXERCISES: There is nothing to practice in this lesson except reproducing a coordinate plane as shown and described. The next lesson introduces plotting and identifying points on the coordinate plane which requires some practice.

INSTRUCTOR'S NOTES:

LESSON 46

AIM/TOPIC(S): Plotting points on the coordinate plane

WARM-UP: Have the student draw a coordinate plane. Draw one yourself, and plot the points shown at left without labeling the coordinates. Elicit the student's ideas on what the "addresses" of the points might be, or how we could identify them. Use any response to lead into the lesson.

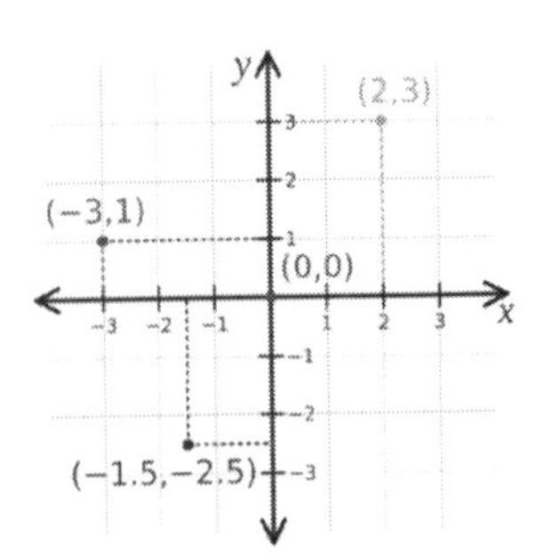

MOTIVATION: This lesson teaches a fundamental aspect of working with a coordinate plane—identifying and plotting points. It is very easy, but for some reason many students get confused, and insist on overly complicating the entire matter.

POINTS TO ELICIT: Revisit the warm-up. There are different ways of tackling this topic. Have the student look at the point (2, 3). Elicit that every point has an "x address" and a "y address." This is similar to a postal address involving a street number and an avenue number. Explain that the address of a point is called its coordinates. We always list the x-coordinate first, followed by the y-coordinate. Demonstrate the notation using an ordered pair in parentheses. Elicit that the point (2, 3) is "aligned" with 2 on the x-axis, and 3 on the y-axis.

The first opportunity for confusion comes when negative points are involved. Examine the point (-3, 1). It is "aligned" with -3 on the x-axis, and 1 on the y-axis. Do not let the student make the matter any more complicated than that.

Some students panic when they see the negative, and assume that they are supposed to list the y-coordinate first, or write the negative coordinate as positive, or similar misconceptions.

Examine the point with decimal coordinates, and elicit that they are correct. Explain to the student that s/he will not likely be expected to plot points with fractional coordinates.

Examine the origin. By our procedure, its coordinates are (0, 0). The origin is our starting point or "home." To get to various points, we always start at the origin, and then "walk" a given number of "blocks" left/right, followed by a given number of "blocks" up/down.

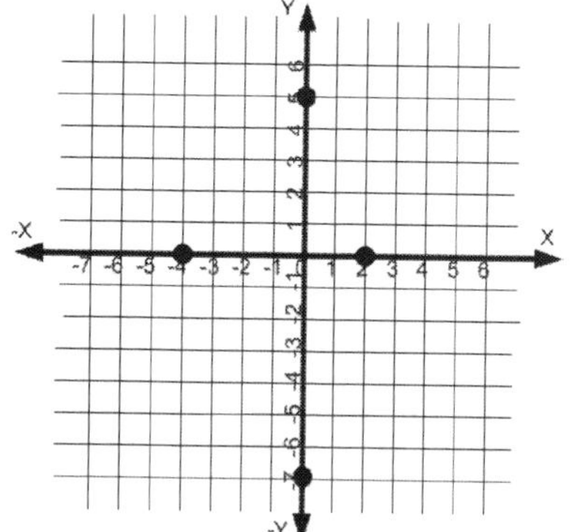

Draw another coordinate plane, plot the points shown, and try to elicit their coordinates. Many students get confused with points on an axis. This may be because it isn't possible to draw dotted lines like we could in the previous diagram. It may also be because students are often not willing to accept 0 as a value.

Elicit that our procedure is still the same. To determine the coordinates of a point, we still start at the origin. We first determine how far left or right to go, which may be 0. We then determine how far up or down to go, which may be 0.

PRACTICE EXERCISES: Keep practicing until the student has no difficulty plotting or identifying points.

INSTRUCTOR'S NOTES:

LESSON 47

AIM/TOPIC(S): Computing the slope between two points

PREREQUISITE KNOWLEDGE: It is essential that the student have no difficulty with signed number arithmetic.

WARM-UP: There is no warm-up for this lesson since the student would have no idea how to proceed. Use the warm-up time to ensure that the last two lessons were mastered, and that the student has the prerequisite skills described above.

MOTIVATION: A common task in coordinate geometry is to compute the slope between two points. It requires a simple formula, and the prerequisite knowledge described above.

POINTS TO ELICIT / DEMONSTRATIVE EXAMPLES:

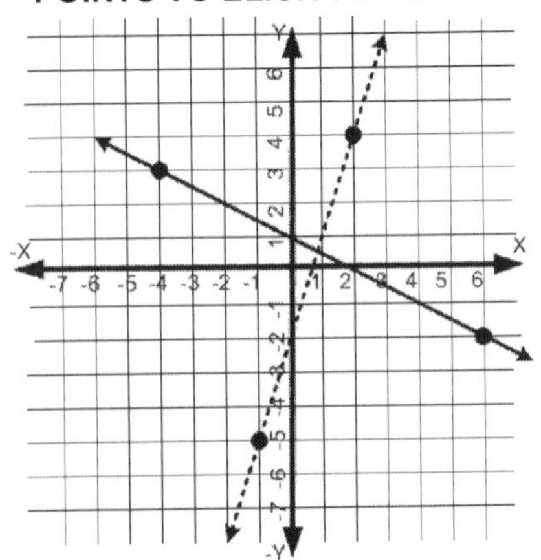

Reproduce the coordinate plane and lines at left. We have a formula which allows us to compute what we call the slope of a line. Informally, the slope tells us a line's steepness and direction of slant.

There are many different versions of the slope formula, all of which mean exactly the same thing. The slope between two points on a line is defined as the "change in y" divided by the "change in x." We could equivalently define it as the vertical change divided by the horizontal change.

Let's compute the slope of the dotted line in the coordinate plane at left. We use the letter *m* to represent slope. To begin, we must pick one of the two points on the line, and call it Point A (or Point 1). It doesn't matter which point we pick. The other point will be Point B (or Point 2). It is essential to stress that once we pick which point will be which, we must be consistent throughout the entire problem.

To compute the vertical change between the two points, we must subtract their y-coordinates. That is how we compute the difference between them. Let's call the uppermost point A, and the lowermost point B. When we subtract the y-coordinates we get 4 – (-5) which is 9. To compute the horizontal change, we must subtract the x-coordinates while maintaining the same point names. We'll compute 2 – (-1) to get 3. The slope is defined as the change in y divided by the change in x, so we must compute 9/3 to get 3. We say that the slope is 3.

Let's find the slope of the other line. If we call the uppermost point A, we can compute the change in y as 3 – (-2) which is 5. The change in x is (-4) – 6 which is -10. Notice how careful we must be with the signs. We must compute the change in y divided by the change in x which is -½, our slope.

PRACTICE EXERCISES: Have the student recompute the slopes of these lines, but reversing points A and B. S/he should get the exact same answers. The only difference is that during the intermediate computations, the signs will be different.

Continue to practice computing the slope between any two points. For now, avoid horizontal and vertical lines which will be covered next. It is fine if the slope works out to be a fraction.

INSTRUCTOR'S NOTES:

LESSON 48

AIM/TOPIC(S): Positive, negative, zero, and undefined slopes

WARM-UP: As with the last lesson, a warm-up for this lesson is difficult since the student will not likely know how to proceed. Use the time for review of the previous lesson.

MOTIVATION: This lesson extends the concept of slope to horizontal and vertical lines. The concepts can be tricky.

POINTS TO ELICIT: Reproduce the coordinate plane and lines below. Examine the horizontal dotted line. Elicit that all points along that line have the same y-coordinate. That coordinate happens to be -2, but that doesn't matter. When we subtract the y-coordinates between any two chosen points on the line, we get -2 – (-2) which is 0. This makes sense since there is no vertical change between the two points.

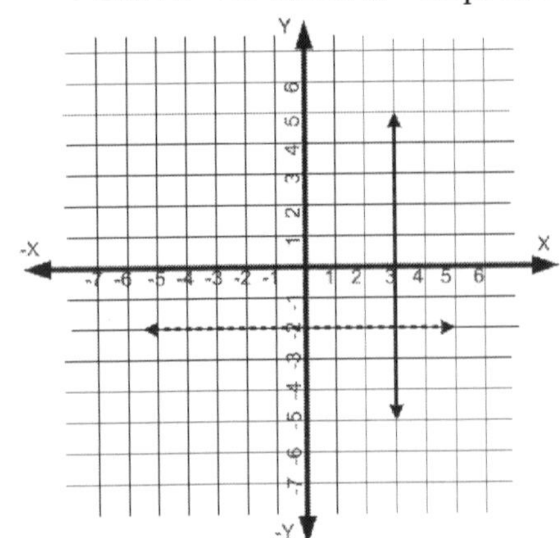

Now let's look at the horizontal change along the line. Elicit that if we pick two points on the line, there will certainly be some horizontal change between them. It doesn't matter how much. The slope formula has us take the vertical change and divide it by the horizontal change. Elicit that we have the situation of $0/n$ which is 0. We say that the slope of a horizontal line is 0. Think of it as driving along a level (flat) road.

Now examine the vertical solid line. All points along the line have the same x-coordinate. When we subtract the x-coordinates between any two chosen points on the line we get 3 – 3 which is 0. This makes sense since there is no horizontal change between the two points.

Let's look at the vertical change along the line. If we pick two points there will certainly be a vertical change between them—it doesn't matter how much. The slope formula has us compute the vertical change divided by the horizontal change.

Elicit that we have the situation of $n/0$ which is undefined. Review the basic math fact that we are "not allowed" to divide by 0. The answer is not 0, it is "undefined." A calculator would generate an error message. We say that the slope of a vertical line is undefined. Think of it as trying to drive straight down the face of a cliff.

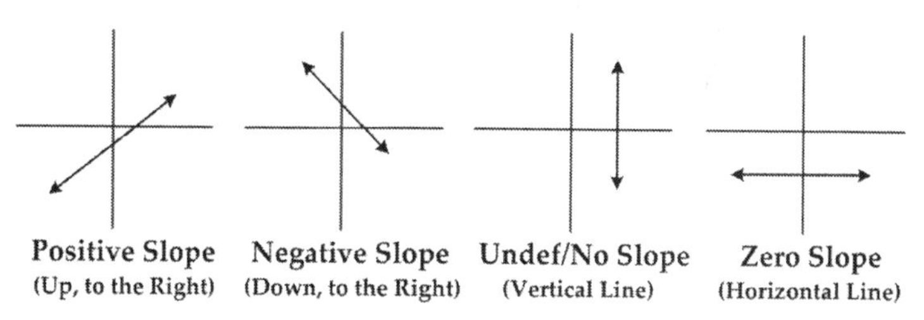

Positive Slope (Up, to the Right) **Negative Slope** (Down, to the Right) **Undef/No Slope** (Vertical Line) **Zero Slope** (Horizontal Line)

After practicing the last lesson, the student should discover that lines with positive slopes run "up and to the right," and lines with negative slopes run "down and to the right." The chart above illustrates what we have learned about slopes.

PRACTICE EXERCISES: Spend time reviewing the concepts of this lesson, as well as the general use of the slope formula.

INSTRUCTOR'S NOTES:

LESSON 49

AIM/TOPIC(S): Y-intercept; Slope-intercept form of a line

WARM-UP: As with the previous lesson, use the warm-up time to review the previous lessons in this chapter.

MOTIVATION: In this very important lesson the student will learn the concept of y-intercept, as well as the slope-intercept ($y = mx + b$) form of a line. Ensure that the lesson is mastered.

POINTS TO ELICIT / DEMONSTRATIVE EXAMPLES:

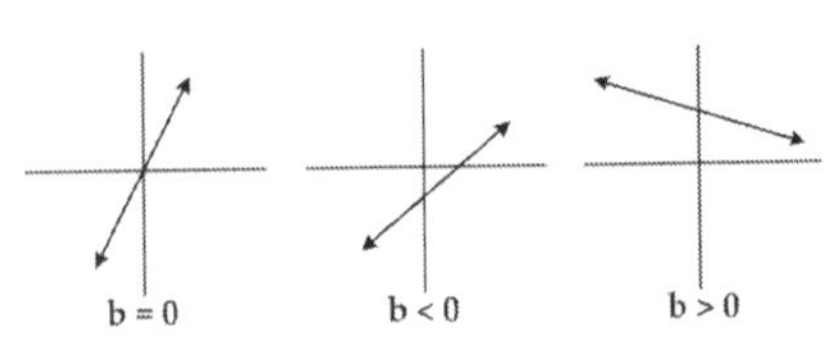

Reproduce the diagrams at left. Explain that in addition to slope, we can determine the y-intercept of a line, denoted b. The y-intercept tells us where the line intersects the y-axis. Once we know the slope and y-intercept, we have a complete picture of what the line looks like on the coordinate plane.

Examine the first line. Ignoring slope, where does the line intersect (cross) the y-axis? Elicit that the answer is 0. We say that the y-intercept of that line is 0. Do the same for the second line. Again, when determining y-intercept, slope is irrelevant. We don't know exactly where the line intersects the y-axis, but we know that it is somewhere in the negative area. We know that $b < 0$. If we examine the third line, we can see that $b > 0$.

Introduce the concept of a linear equation in slope-intercept form ($y = mx + b$) using the coordinate plane at right as a model. The line has an equation of $y = 3x + 2$. Using our equation model, we see that the slope is 3, and y-intercept is 2.

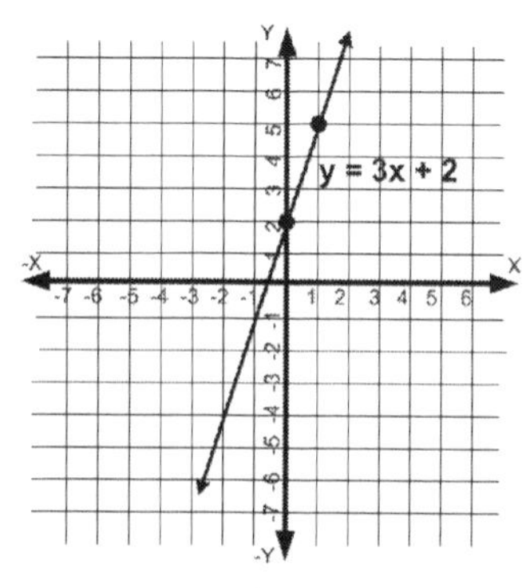

That is all we need to know to graph the line. One of the points will be (0, 2) which is the y-intercept. The slope of positive 3 tells us that the line will slant up and to the right. If we convert the slope of 3 into the equivalent 3/1, we can see that this line has a vertical change of 3 for every horizontal change of 1. This is sometimes described as a rise of 3 for a run of 1. Have the student examine the diagram until this is clear.

Here is another example. The y-intercept is -3, and the slope is -2. The "y=mx+b" form of a line specifically calls for a plus sign, so the fact that this line's equation has "minus 3" means the y-intercept is negative. One of the points on the line is (0, -3) which is the y-intercept. If we convert the slope of -2 into -2/1, we see that instead of "rising" 2 squares for every square we move to the right, we will instead "fall" two. This results in the downward slant which was expected from a negative slope.

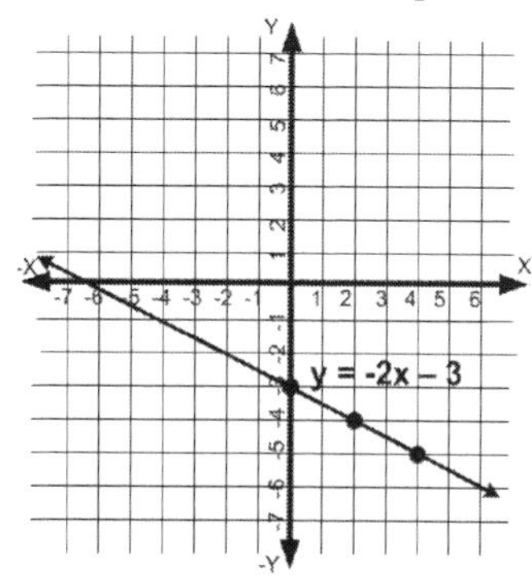

PRACTICE EXERCISES: In later math the student will do much more with these concepts. Just review them and practice with similar exercises. The main book has more examples.

INSTRUCTOR'S NOTES:

LESSON 50

AIM/TOPIC(S): Converting from standard to y=mx+b form

WARM-UP: This lesson relies heavily upon basic algebra skills. Ensure this is not a point of concern. Ask the student to state the slope and y-intercept of the line represented by the equation $6x + 2y = 8$. Use any response to lead into the lesson.

MOTIVATION: In this short lesson we will practice converting from the standard form of a line to the slope-intercept form.

POINTS TO ELICIT / DEMONSTRATIVE EXAMPLES:
Revisit the warm-up. Most students will come up with answers without actually doing any algebra. Elicit that we cannot visually determine the slope and y-intercept of a linear equation until it has been converted into slope-intercept form.

The equation in the warm-up is in standard form. We can use basic algebraic techniques to convert it into slope-intercept form. "Move" the 6x over to the right by subtracting 6x from each side. That gives us $2y = 8 - 6x$. We need y by itself on the left so we must divide both sides by 2, giving us $y = 4 - 3x$. Notice how on the right we distributed the division over the subtraction. We need the x term to come first, so we must rewrite the equation as $y = -3x + 4$. Now that the equation is in y=mx+b form, we can easily see that m is -3 and b is 4.

PRACTICE EXERCISES: For additional practice, have the student convert $5x - 3y = -6$ into slope-intercept form. We get $y = (5/3)x + 2$. $m = 5/3$ and $b = 2$. Many students have difficulty with this lesson. Continue to review basic algebra skills.

INSTRUCTOR'S NOTES:

Assessing and Working with Students Who Are Struggling with Basic Algebra and Geometry

This advice in this section is almost identical to that which was offered in the basic math lesson plan book in this series. It is repeated here because it is even more relevant and applicable to students are who are studying algebra and geometry.

It is imperative that you have faith in your students' ability to be successful in their math goals. Students of all ages can easily tell if you are not. Do not take on any client for whom you feel you cannot be of benefit. This involves learning to become honest with yourself as well as the client.

The biggest issue that you will face is students who are very far behind in math, but want to catch up by taking very few sessions over a very short amount of time. It is also common for students (and their parents if applicable) to have no idea just how far behind they are in math. Remember that if a student gets passing grades on exams, and gets promoted from one grade to the next, neither the student nor his/her parents have any idea that anything could possibly be wrong.

Another issue you will face is students who are not willing to work on material which they consider to be beneath them, or which they feel they have already mastered. It is important to assess all prospective students before taking them on as clients. It is strongly suggested that you not charge any money to perform such an assessment or discuss its results.

An assessment need not be written nor formal unless insisted upon by a student's parents which is fairly common. If you are a skilled instructor, it should not be hard to verbally spot-check the student to determine his/her math weaknesses. For example, if you ask a high school student to add ½ + ⅓, and give them paper and time in which to do so, a common response will be, "Don't we do something with the denominators? Wow, that was like five years ago!"

In such a case it would be essential to explain to the student (and his/her parents if applicable) that s/he cannot begin to work with algebraic fractions if s/he can't handle simple numeric ones. This of course extends to other basic math topics such as signed number arithmetic which is a prerequisite skill for algebra. Obviously be delicate and compassionate, but the point must be made if you are to be considered ethical.

It is important to be honest with your students (or privately with their parents if applicable) in regards to their goals and expectations. If a student is very far behind in math, but only has time or money for ten sessions with you, do not smile at him/her and say, "Sure! You can do it!" That only works in Hollywood movies. Instead, discuss the situation realistically while being positive and encouraging.

Contact me via my website if you have questions about the lessons in this book, or would like to discuss the needs of a student that you are working with. Now get out there and make a difference for some students! ☺

Instructor's Notes on Students' Progress

Tutors or group instructors are encouraged to use the space on this and the following pages to keep track of where they left off with each student at the end of each session. Students and parents are always impressed by an instructor who confidently begins each session by asserting, "Last time we left off at..." Be sure to include what lesson number you were up to, whether or not you completed it, and any concepts which were a point of concern for the student.

Don't assume that the student remembers any or all of the material from the last lesson, or that s/he completed any studying or assigned practice exercises. Don't assume that the student will be sophisticated enough to start the lesson by asking any questions that s/he may have. Begin each lesson with, at the very minimum, an informal verbal quiz to assess how much material from the last session was retained.

STUDENT'S NAME: ______________________________

NOTES: ______________________________

____

____

____

____

____

____

____

STUDENT'S NAME: ________________________________
NOTES: ____

STUDENT'S NAME: ________________________________
NOTES: ____

STUDENT'S NAME: ________________________________
NOTES: ____

STUDENT'S NAME: ______________________________
NOTES: ______________________________

STUDENT'S NAME: ______________________________
NOTES: ______________________________

STUDENT'S NAME: ______________________________
NOTES: ______________________________

STUDENT'S NAME: ______________________________
NOTES: ______________________________

STUDENT'S NAME: ______________________________
NOTES: ______________________________

STUDENT'S NAME: ______________________________
NOTES: ______________________________

Acknowledgements

I would like to thank my fiancée Carlee Dise for proofreading and providing suggestions for this and all of the other self-published books in the *Math Made a Bit Easier* series, and for her encouragement and support through the entire project.

I would also like to thank the online bookseller which at press time is the exclusive seller of my books, and its subsidiary which provides authors with the means to easily and cost-effectively self-publish their books. For contractual reasons, the company does not currently permit its name or the name of its subsidiary to be mentioned in any self-published books.

About the Author

Larry Zafran was born and raised in Queens, NY where he tutored and taught math in public and private schools. He has a Bachelors Degree in Computer Science from Queens College where he graduated with highest honors, and has earned most of the credits toward a Masters in Secondary Math Education.

He is a dedicated student of the piano, and the leader of a large and active group of board game players which focuses on abstract strategy games from Europe.

He presently lives in Cary, NC where he works as an independent math tutor, writer, and webmaster.

Companion Website for More Help

For free support related to this or any of the author's other math books, please visit the companion website below.

www.MathWithLarry.com